A-level
In a Week

Chemistry

Year 2

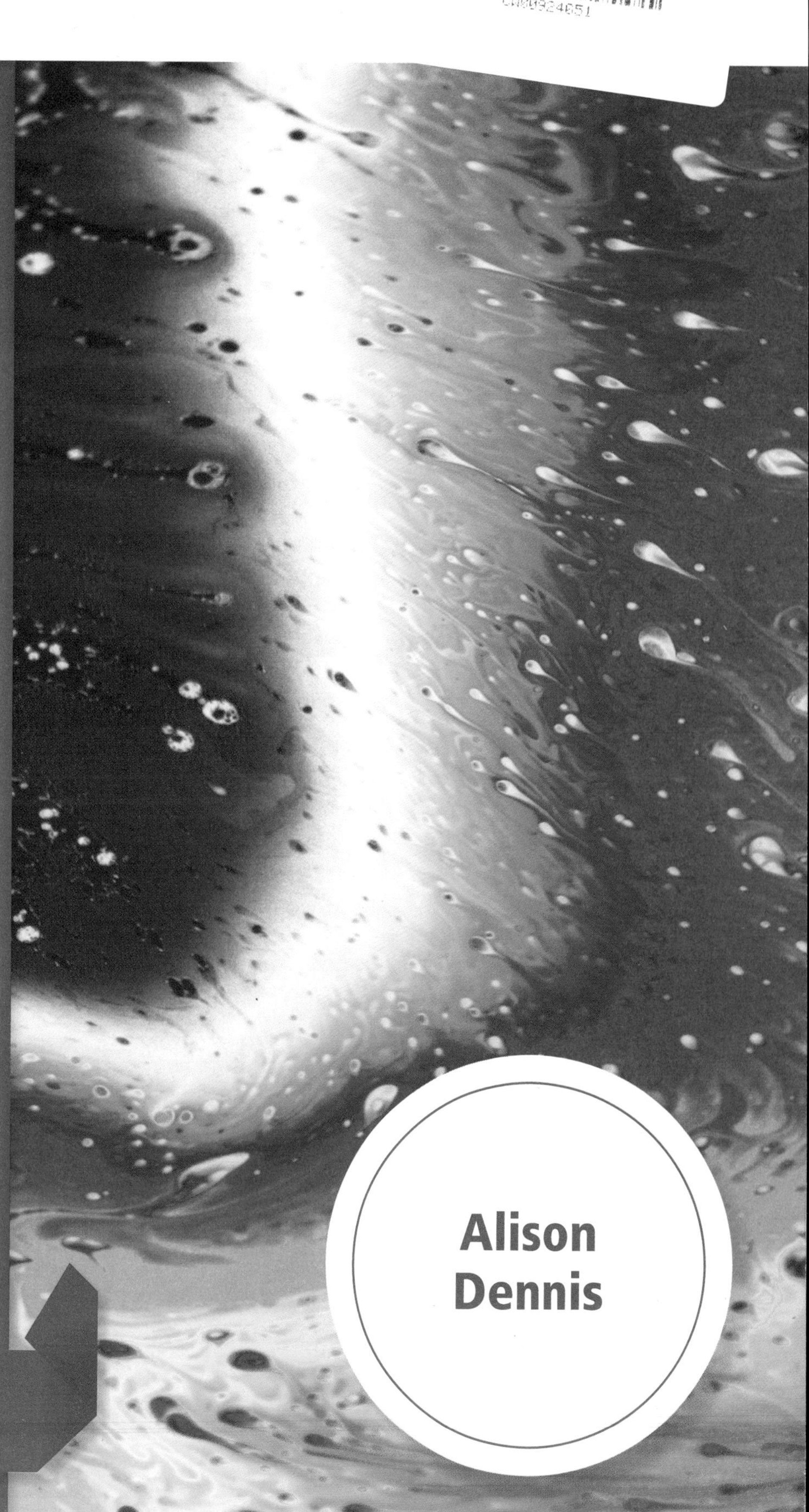

Alison Dennis

CONTENTS

DAY 1 60 Minutes

Equilibria

Equilibrium Constant K_c

The equilibrium expression is written from the reaction equation, e.g.

$$3H_2(g) + N_2(g) \rightleftharpoons 2NH_3(g)$$

Each reactant and product is written as a concentration by using square brackets.	$[NH_3(g)]$ $[H_2(g)]$ $[N_2(g)]$
Each concentration is raised to the power of its coefficient in the reaction equation.	$2NH_3(g) = [NH_3(g)]^2$ $3H_2(g) = [H_2(g)]^3$ $N_2(g) = [N_2(g)]$
The products are multiplied together on the top of the equation and the reactants multiplied together on the bottom of the equation.	$\frac{[NH_3(g)]^2}{[H_2(g)]^3 [N_2(g)]}$

Note: The concentration of a solid in an equilibrium can be considered to be 1, so is not included in the equilibrium expression.

Calculating the Equilibrium Constant from Experimental Results

The concentration of any substance in the equilibrium mixture is found by adding together the initial concentration and the change in concentration for that substance, e.g.

0.10 moles of $C_2H_5OH(l)$ were mixed with 0.20 moles of $CH_3COOH(l)$. A concentrated acid catalyst was added and the total volume of the mixture was 125 cm^3. The reaction was left to come to equilibrium.

$$C_2H_5OH(l) + CH_3COOH(l) \rightleftharpoons C_2H_5OCOCH_3(l) + H_2O(l)$$

The number of moles of CH_3COOH at equilibrium was calculated by titrating against sodium hydroxide and found to be 0.115. Calculate the value of K_c for the reaction at this temperature.

1. Write the equilibrium expression for the reaction.

$$K_c = \frac{[C_2H_3OCOCH_3(l)][H_2O(l)]}{[C_2H_5OH(l)][CH_3COOH(l)]}$$

2. Draw up a table as below and write in the values that are known.

   a. starting moles of each substance

   b. moles at equilibrium

   c. change in moles from start to equilibrium

   d. concentration ($mol\,dm^{-3}$)

	C_2H_5OH	CH_3COOH	$C_2H_5OCOCH_3$	H_2O
a	0.1	0.2	0	0
b		0.115		
c				
d				

3. Calculate the change in moles for each substance.

   Change in moles of CH_3COOH = starting moles – equilibrium moles = 0.2 – 0.115 = 0.085 moles acid. These moles of CH_3COOH were converted to ester.

   From the reaction equation every 1 mole of CH_3COOH used makes 1 mole ester and 1 mole water. Change in moles of ester and water = +0.085 moles

   Every mole of ester made used up one mole of alcohol. Change in moles of alcohol = –0.085 moles

4. Add the change in moles to the starting moles to calculate the moles of each substance at equilibrium.

	C_2H_5OH	CH_3COOH	$C_2H_5OCOCH_3$	H_2O
a	0.1	0.2	0	0
b	0.015	0.115	0.085	0.085
c	–0.085	–0.085	+0.085	+0.085
d	0.015 ÷ 0.125 = 0.12	0.115 ÷ 0.125 = 0.92	0.085 ÷ 0.125 = 0.68	0.085 ÷ 0.125 = 0.68

5. Use the total volume of the reaction mixture and the moles of each reactant to calculate the concentration of each reactant.

$$\text{Conc} = \frac{\text{moles}}{\text{volume}}$$

6. Insert the concentration values into the equilibrium expression.

$$K_c = \frac{0.68 \times 0.68}{0.12 \times 0.92} = 4.19$$

7. Calculate the units of K_c by writing all the units of concentration into the equilibrium expression and cancelling down.

$$\text{units } K_c = \frac{\cancel{\text{mol dm}^{-3}} \times \cancel{\text{mol dm}^{-3}}}{\cancel{\text{mol dm}^{-3}} \times \cancel{\text{mol dm}^{-3}}}$$

no units

No Change in K_c with a Catalyst or Changed Concentration

Changing the **concentration** of any reactant and using a **catalyst** have **no effect** on the value of K_c.

Changing Temperature Changes K_c

	Exothermic reaction	Endothermic reaction
Increasing temperature	decreases K_c	increases K_c
Decreasing temperature	increases K_c	decreases K_c

The value of K_c is always quoted for a specific temperature.

Equilibrium Constant K_p

The equilibrium constant for a reaction involving gases can be calculated as K_p.

K_p is written in the same way as K_c but using partial pressures of gases instead of concentrations. Partial pressure can be written as, for example, pO_2 or Po_2 (no brackets).

The Mole Fraction

$$\text{Mole fraction of a gas in a mixture} = \frac{\text{moles of gas}}{\text{total moles of all gases}}$$

Partial Pressures of Gases

In a mixture of gases, the pressure exerted by each individual gas is known as the partial pressure of that gas. The total pressure is the sum of the partial pressures of the gases in the mixture.

Partial pressure of a gas = mole fraction × total pressure

Calculating K_p from Data

1. Write the equilibrium expression for K_p.
2. Calculate the number of moles of each gas at equilibrium.
3. Add all the moles of gas to find the total moles of gas at equilibrium.
4. Find the mole fraction of each gas by dividing the moles of each gas at equilibrium by the total moles.
5. Find the partial pressure of each gas by multiplying the mole fraction by the total pressure at equilibrium.
6. Insert the partial pressures for each gas in the expression for K_p.
7. Work out the units of K_p by inserting the units into the K_p expression and cancelling down.

Example

$$3H_2(g) + N_2(g) \rightleftharpoons 2NH_3(g)$$

60.0 moles of $H_2(g)$ and 40.0 moles $N_2(g)$ were allowed to come to equilibrium at 150 atm. At equilibrium 50% of the hydrogen was converted to ammonia. Calculate the value of K_p.

1. $K_p = \frac{pNH_3^2}{pH_2^3 \times pN_2}$

2. 50% H_2 was converted to NH_3.
   Remaining moles of H_2 = 60 × 50% = **30** moles

   Each mole of H_2 reacts with $\frac{1}{3}$ mole N_2.

   Moles N_2 used in reaction = 30 × $\frac{1}{3}$ = 10 moles

   Moles N_2 remaining at equilibrium = 40 − 10
   = **30** moles

   Each mole N_2 used forms 2 moles NH_3.
   Moles NH_3 at equilibrium = 10 × 2 = **20** moles

3. Total moles of gas at equilibrium
   = 20 + 30 + 30 = 80 moles

4. Mole fraction: $H_2 = \frac{30}{80} = 0.375$

   $N_2 = \frac{30}{80} = 0.375$ $NH_3 = \frac{20}{80} = 0.250$

5. Pressure = 150 atm partial pressures:
$H_2 = 0.375 \times 150 = 56.25$ atm
$N_2 = 0.375 \times 150 = 56.25$ atm
$NH_3 = 0.250 \times 150 = 37.5$ atm

6. $K_p = \frac{37.5^2}{56.25^3 \times 56.25} = 1.40 \times 10^{-4}$

7. $\frac{\cancel{atm} \times \cancel{atm}}{\cancel{atm} \times \cancel{atm} \times atm \times atm} = atm^{-2}$

Solubility Equilibria, K_{sp}

A sparingly soluble solid dissolved in water to form a saturated solution is in equilibrium.

$$AgCl\,(s) \rightleftharpoons Ag^+(aq) + Cl^-(aq)$$

$$K_c = \frac{[Ag^+(aq)][Cl^-(aq)]}{[AgCl(s)]}$$

Adding more AgCl(s) to the solution does not increase the concentration of Ag^+(aq) and Cl^-(aq) because the solution is already saturated. Since AgCl is a solid it can be left out of the equilibrium expression. In this case K_c is given the name K_{sp}, the solubility product.

$$K_{sp} = [Ag^+(aq)]\,[Cl^-(aq)] = 2 \times 10^{-10}\ mol^2\,dm^{-6}$$

If the value of K_{sp} is exceeded by adding **either** Ag^+(aq) **or** Cl^-(aq) to a saturated solution of AgCl then more solid will precipitate, e.g. adding $AgNO_3$(aq) or NaCl(aq) will cause AgCl(s) to precipitate.

Solubility products only apply to saturated solutions.

Determining Solubility Products

Make up a saturated solution of the sparingly soluble solid. Remove an accurately known volume. Analyse the sample to find the concentration of one of the ions, e.g. by titration. Use the formula of the compound to calculate the concentration of the other ion. Insert the concentration values into the expression for K_{sp}.

QUICK TEST

1. Write the expression for K_c for the reaction $ClO(g) + NO_2(g) \rightleftharpoons ClONO_2(g)$
2. What are the units of K_c for this reaction?
3. Calculate the equilibrium concentration of ClO if the starting quantity of ClO(g) was 5 moles, the starting moles of NO_2(g) was 6 moles and at equilibrium there were 2 moles NO_2(g) in a total volume of 2.5 dm^3.
4. Write the equilibrium expression K_p for $H_2(g) + Cl_2(g) \rightleftharpoons 2HCl(g)$
5. Calculate the units of K_p for this reaction.
6. Write the expression for K_{sp} for the reaction $PbI_2(s) \rightleftharpoons Pb^{2+}(aq) + 2I^-(aq)$

PRACTICE QUESTIONS

1. Ethene reacts with steam: $CH_2{=}CH_2 + H_2O \rightleftharpoons CH_3CH_2OH$

The units of K_c for this reaction are **[1 mark]**

A $mol\ dm^{-3}$ ☐

B $mol^2\ dm^{-3}$ ☐

C $mol^{-1}\ dm^3$ ☐

D no units. ☐

2. Carbon dioxide and nitrogen oxide react together: $CO_2(g) + NO(g) \rightleftharpoons CO(g) + NO_2(g)$

The effect of using a catalyst is to **[1 mark]**

A increase the concentration of products ☐

B decrease the value of K_c ☐

C increase the value of K_c ☐

D have no effect on the value of K_c. ☐

3. A sample of 1.2×10^{-1} moles of ethanoic acid was mixed with 1.8×10^{-3} moles of ethanol and left to come to equilibrium.

$$C_2H_5OH(l) + CH_3COOH(l) \rightleftharpoons C_2H_5OCOCH_3(l) + H_2O(l)$$

The resulting mixture reacted with 24.0 cm^3 of 1.00 mol dm^{-3} KOH(aq). The total equilibrium volume was 200 cm^3.

a) Write the expression for K_c for this equilibrium. **[1 mark]**

b) Calculate the concentration of ethanoic acid at equilibrium. **[2 marks]**

c) Calculate the value and units of K_c at this temperature. **[4 marks]**

4. For the equilibrium:

$$PCl_5(g) \rightleftharpoons PCl_3(g) + Cl_2(g) \qquad \Delta H^{\ominus} +87 \text{ kJ mol}$$

When 9.00 moles $PCl_5(g)$ was allowed to come to equilibrium at 2.63 atm and 500 K there were found to be 3.60 moles of Cl_2 in the equilibrium mixture.

a) Calculate the moles of PCl_3 and PCl_5 at equilibrium. **[1 mark]**

b) Calculate the value for K_p under these conditions. Include units in your answer. **[2 marks]**

c) What would be the effect on K_p if the temperature was changed to 300 K and why? **[2 marks]**

5. Barium meals are used to investigate problems with the gastrointestinal tract. The primary ingredient of a barium meal is $BaSO_4(s)$, a sparingly soluble salt.

a) Write the expression for K_{sp} for barium sulfate. **[1 mark]**

b) Explain why this expression does not include $BaSO_4(s)$. **[1 mark]**

c) What would be observed if magnesium sulfate solution was added to a saturated solution of barium sulfate? **[1 mark]**

Acids and pH

Brønsted–Lowry Acids

Brønsted–Lowry acids are proton (H^+) donors, they dissociate into an H^+ and a negative ion, e.g.

$$HCl(aq) \longrightarrow H^+(aq) + Cl^-(aq)$$

In simple solution H^+ is donated to water.

$$HCl(aq) + H_2O(l) \longrightarrow H_3O^+(aq) + Cl^-(aq)$$

$$CH_3COOH\ (aq) + H_2O(l) \rightleftharpoons H_3O^+(aq) + CH_3COO^-(aq)$$

$H_3O^+(aq)$ is called a hydronium ion and is often written $H^+(aq)$.

Brønsted–Lowry Bases

Brønsted–Lowry bases are proton acceptors. They bond with H^+ from an acid.

$$NH_3(aq) + H^+(aq) \longrightarrow NH_4^+\ (aq)$$

Conjugate Acid–Base Pairs

The ionisation of an acid is reversible.

$$H_3O^+(aq) + CH_3COO^-(aq) \rightleftharpoons CH_3COOH\ (aq) + H_2O(l)$$

The anion of the acid, CH_3COO^-, accepts a proton to reform the acid CH_3COOH. CH_3COO^- is a proton acceptor. It is called the **conjugate base** of CH_3COOH.

CH_3COOH/CH_3COO^- are a conjugate acid–base pair.

A base that has accepted a proton is able to become a proton donor.

$$NH_4^+(aq) \longrightarrow NH_3(aq) + H^+(aq)$$

NH_4^+ is the **conjugate acid** of NH_3.

NH_4^+/NH_3 are a conjugate acid–base pair.

Mono, Di and Tribasic Acids (also called Mono, Di and Triprotic Acids)

A monobasic acid can donate one proton, a dibasic two protons and a tribasic three protons, e.g.

Monobasic $HNO_3 \longrightarrow H^+ + NO_3^-$

Dibasic $H_2SO_4 \longrightarrow 2H^+ + SO_4^{2-}$

Tribasic $H_3PO_4 \longrightarrow 3H^+ + PO_4^{3-}$

Reactions of Acids

Reactions of acids involve H^+ and can be written as ionic equations. $H^+(aq)$ can be from any acid.

Acid + alkali (soluble base):

Acid + alkali ⟶ salt + water

$$H^+(aq) + OH^-(aq) \longrightarrow H_2O(l)$$

Acid + insoluble base:

Acid + metal oxide/hydroxide ⟶ salt + water

$$2H^+(aq) + M(OH)_2(s) \longrightarrow M^{2+}(aq) + 2H_2O(l)$$

$$2H^+(aq) + MO(s) \longrightarrow M^{2+}(aq) + H_2O(l)$$

Acid + carbonate:

Acid + carbonate/hydrogen carbonate ⟶ salt + water + carbon dioxide

$$2H^+(aq) + CO_3^{2-}(aq) \longrightarrow H_2O(l) + CO_2(g)$$

$$H^+(aq) + HCO_3^-(aq) \longrightarrow H_2O(l) + CO_2(g)$$

Note: A carbonate is also a base since it accepts protons.

Strong and Weak Acids and Bases

Strong acids and bases fully ionise in solution.

$$HCl\ (aq) \longrightarrow H^+(aq) + Cl^-(aq)$$

$$NaOH(aq) \longrightarrow Na^+(aq) + OH^-(aq)$$

Weak acids and bases only partially ionise and so form equilibria.

$$CH_3COOH(aq) \rightleftharpoons H^+(aq) + CH_3COO^-(aq)$$

$$NH_3(aq) + H^+(aq) \rightleftharpoons NH_4^+(aq)$$

K_a Shows the Strength of an Acid

The extent to which a weak acid ionises is shown by the acid dissociation constant, K_a.

$HA(aq) \rightleftharpoons H^+(aq) + A^-(aq)$ $\qquad K_a = \frac{[H^+(aq)][A^-(aq)]}{[HA(aq)]}$

The larger K_a, the stronger the acid.

pK_a is $-lgK_a$

The value of K_a is commonly very small and can vary widely. A more convenient way of describing K_a is as a negative logarithm: $pK_a = -logK_a$ $\qquad K_a = 10^{-pK_a}$

	K_a	pK_a
Hydrocyanic acid	4.9×10^{-10}	9.3
Nitrous acid	4.7×10^{-4}	3.3

The higher pK_a, the weaker the acid.

Enthalpy Change of Neutralisation $\Delta H^{\ominus}_{neut}$

The enthalpy change of neutralisation is the enthalpy change when one mole of water is produced from the neutralisation of an acid under standard conditions.

When a strong, fully dissociated acid is neutralised, all H^+ ions are already in solution and the reaction consists of forming the bond between H^+ and OH^-. Making bonds releases energy. $\Delta H^{\ominus}_{neut}$ is exothermic.

When a weak acid is neutralised only a small percentage of H^+ is already in solution and the remainder is formed by breaking covalent bonds. Breaking bonds requires energy. As a result the overall energy released is less for weak acids than for strong acids. $\Delta H^{\ominus}_{neut}$ is more negative for strong acids than for weak acids.

pH of Strong Acids

All acidic solutions contain hydrogen ions. The acidic properties depend on the concentration of hydrogen ions, which is given by the pH.

pH is the negative logarithm of the concentration of hydrogen ions in solution.

$pH = -log[H^+]$

The $[H^+]$ of strong monobasic acids = concentration of the acid, e.g. $0.1\ mol\ dm^{-3}\ HNO_3 \longrightarrow 0.1\ mol\ dm^{-3}\ H^+$.

$[H^+] = 0.1\ mol\ dm^{-3}$ or $1 \times 10^{-1}\ mol\ dm^{-3}$

$log[H^+] = -1$ $\qquad$ negative log/pH = 1

$[H^+]$ $mol\ dm^{-3}$	or	pH
0.1	1×10^{-1}	pH 1
0.01	1×10^{-2}	pH 2
0.001	1×10^{-3}	pH 3

pH of $0.025\ mol\ dm^{-3}$ HCl $\qquad$ $log\ 0.025 = -1.60$

$-log = 1.60$ $\quad$ pH = 1.60

H_2SO_4 is a dibasic acid: there are two dissociations.

$H_2SO_4 \longrightarrow H^+ + HSO_4^-$ $\qquad HSO_4^- \rightleftharpoons H^+ + SO_4^{2-}$

Although the second dissociation is not as strong, for pH calculations at A-level, it is assumed that

$1\ mol\ dm^{-3}\ H_2SO_4 \longrightarrow 2\ mol\ dm^{-3}\ H^+$

pH of $0.01\ mol\ dm^{-3}\ H_2SO_4$

$[H^+] = 0.02\ mol\ dm^{-3}$ $\quad -log[H^+] = 1.7$ $\quad pH = 1.7$

K_w, the Ion/Ionic Product of Water

Water ionises to a small extent.

$$H_2O \rightleftharpoons H^+ + OH^-$$

The equilibrium expression for this reaction is

$K_a = \frac{[H^+(aq)][OH^-(aq)]}{[H_2O(l)]}$

The concentration of water is constant, $[H_2O] = 1$
Therefore $K_a = [H^+(aq)][OH^-(aq)]$

This is known as K_w, the ion/ionic product of water.

For pure water at 298 K, $K_w = 1 \times 10^{-14}\ mol^2\ dm^{-6}$

In pure water $[H^+] = [OH^-]$
$[H^+] = \sqrt{1 \times 10^{-14}} = 10^{-7}$ $\qquad$ pH = 7

pK_w is the negative logarithm of K_w $\qquad pK_w = 14$

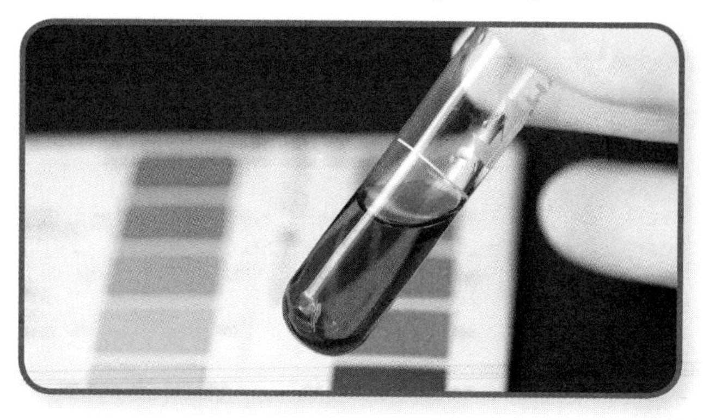

pH of Strong Bases

The concentration of hydroxide ions in a strong base can be calculated from the concentration of the base.

$$KOH \longrightarrow K^+ + OH^-$$
$$0.1\ mol\ dm^{-3}\ KOH \longrightarrow 0.1\ mol\ dm^{-3}\ OH^-$$

Assuming that water makes no contribution to the $[OH^-]$ and that $K_w = [H^+(aq)][OH^-(aq)] = 1.0 \times 10^{-14}$:

For 0.1 mol dm^{-3} OH^-

$1 \times 10^{-14} = [H^+][0.1]$ $[H^+] = 1 \times 10^{-14} \div 0.1$

$[H^+] = 1 \times 10^{-13}$ pH = 13

To find the pH of a strong base at 298 K:

Find $[H^+]$ by dividing 10^{-14} by the concentration of hydroxide ions. $pH = -\log[H^+]$

Alternative method: pOH is the negative logarithm of the concentration of hydroxide ions. $pOH = -\log[OH^-]$

$K_w = [H^+][OH^-]$ $pK_w = pH + pOH$

For water at 298 K $14 = 7 + 7$

$pH = pK_w - pOH$ at 298 K $pH = 14 - pOH$

For 0.1 mol dm^{-3} KOH $pOH = 1$ $pH = 14 - 1 = 13$

The Changes in pH When an Acid Is Diluted

Strong Acids

When a strong acid is diluted the concentration of H^+ is diluted in the same ratio.

[HCl] mol dm^{-3}	$[H^+]$ mol dm^{-3}	pH
0.100	0.10	1.00
0.010	0.010	2.00
0.001	0.001	3.00
0.0001	0.0001	4.00

A tenfold decrease in the acid concentration gives an increase of 1 pH unit. A hundredfold dilution gives an increase of 2 pH units.

Weak Acids

When a weak acid is diluted it affects the position of equilibrium. As more water is added the equilibrium moves to the right and a higher percentage of the acid becomes dissociated.

$$HA(aq) + H_2O(l) \rightleftharpoons H_3O^+(aq) + A^-(aq)$$

The concentration of H^+ decreases but there is some compensation because more H^+ is released from HA.

$[CH_3COOH]$ mol dm^{-3}	$[H^+] \times 10^{-3}$ mol dm^{-3}	pH
0.100	0.00130	2.88
0.01	0.000412	3.39
0.001	0.000130	3.88
0.0001	0.00000412	4.38

At 0.1 mol dm^{-3} ethanoic acid is 1.3% dissociated.

At 0.0001 mol dm^{-3} ethanoic acid is 41% dissociated.

A tenfold dilution gives an increase of less than one pH unit. A hundredfold dilution gives an increase of 1 pH unit.

QUICK TEST

1. $HCOOH(aq) \rightleftharpoons H^+(aq) + HCOO^-(aq)$
   What is the conjugate base of HCOOH?
2. Identify the conjugate acid–base pairs.
   $HClO(aq) + H_2O(l) \rightleftharpoons H_3O^+(aq) + ClO^-(aq)$
3. Name and write the formula of a dibasic acid.
4. Write the ionic equation for the reaction between potassium hydroxide and nitric acid.
5. Which is the stronger acid, chloric acid pK_a 7.4 or benzoic acid pK_a 4.2?
6. What is the concentration of hydroxide ions in pure water at 298 K?

PRACTICE QUESTIONS

1. Sulfuric acid is a dibasic acid: $H_2SO_4(aq) + H_2O(l) \rightleftharpoons HSO_4^-(aq) + H_3O^+(aq)$
Which pair of statements is true? **[1 mark]**

A	H_2SO_4 is the acid.	HSO_4^- is the conjugate base.	☐
B	H_2SO_4 is the acid.	H_3O^+ is the conjugate base.	☐
C	H_2O is the acid.	HSO_4^- is the conjugate base.	☐
D	H_2O is the acid.	H_3O^+ is the conjugate base.	☐

2. Copper carbonate is an insoluble base. Which represents the ionic equation for the reaction of copper carbonate with hydrochloric acid? **[1 mark]**

A $CuCO_3(s) + 2HCl(g) \longrightarrow CuCl_2(aq) + CO_2(g) + H_2O(l)$ ☐

B $CO_3^{2-}(aq) + 2H^+(aq) \longrightarrow 2Cl^-(aq) + CO_2(g) + H_2O(l)$ ☐

C $CuCO_3(s) + 2H^+(aq) \longrightarrow Cu^{2+}(aq) + CO_2(g) + H_2O(l)$ ☐

D $CuCO_3(s) + 2HCl(aq) \longrightarrow CuCl_2(s) + CO_2(g) + H_2O(l)$ ☐

3. The table gives information about some acids.

Acid	**Formula**	**pK_a**
nitric acid	HNO_3	<1
methanoic acid	HCOOH	3.8
benzoic acid	C_6H_5COOH	4.2
propanoic acid	C_2H_5COOH	4.9
chloric(I) acid	HClO	7.4

a) Which is the strongest acid? **[1 mark]**

b) Why are no units given for pK_a? **[1 mark]**

c) What is the value of K_a for benzoic acid? **[1 mark]**

d) Write an equation for the dissociation of aqueous propanoic acid and identify the conjugate acid and base pairs. **[3 marks]**

e) Write the expression for K_a for chloric(I) acid. **[2 marks]**

4. Around 60% of olives are prepared by fermentation. The olives are soaked for several hours in lye, a solution of 0.5 mol dm^{-3} NaOH. The NaOH penetrates part way through the olives making them less bitter. The olives are then washed and placed in brine, NaCl(aq), and some of the NaOH from the olives leaches out into the brine. The pH of the mixture is adjusted with citric acid, a naturally occurring tribasic acid, and the olives are left to ferment.

The ion product of water $K_w = 1.0 \times 10^{-14}$ mol^2 dm^{-6} 298 K

a) Explain what is meant by the statement that citric acid is a tribasic acid. **[1 mark]**

b) Calculate the pH of a sample of the lye. **[2 marks]**

c) If the NaOH that has leached into the brine is 1000 times more dilute than the original lye solution, by how many units will the pH change? **[1 mark]**

Weak Acids, pH and Titrations

A weak acid only partially dissociates so the $[H^+(aq)]$ is much lower than the concentration of acid.

When K_a and concentration of acid are known $[H^+]$ can be calculated by making two assumptions.

Assumption 1: The concentration of acid at equilibrium is the same as the starting concentration of the acid. A weak acid is only slightly dissociated so the loss in concentration due to the equilibrium is insignificant.

Assumption 2: The equilibrium concentration of H^+ and the conjugate base of the acid are the same.
$HA \rightleftharpoons H^+ + A^-$ $[H^+] = [A^-]$
This assumes there is no contribution to $[H^+]$ from the dissociation of water.

Finding the pH of a Weak Acid

If $K_a = \frac{[H^+][A^-]}{[HA]}$ $K_a = \frac{[H^+][H^+]}{\text{[initial concentration of acid]}}$

$K_a \times$ initial concentration of acid $= [H^+]^2$

$\sqrt{(K_a \times \text{initial concentration of acid})} = [H^+]$

$pH = -\log [H^+]$

Example
What is the pH of 0.1 mol dm^{-3} ethanoic acid?

K_a for ethanoic acid $= 1.7 \times 10^{-5}$ mol dm^{-3}

$K_a \times$ concentration of acid $= [H^+]^2$

$1.7 \times 10^{-5} \times 0.1 = [H^+]^2 = 1.7 \times 10^{-6}$

$[H^+] = \sqrt{1.7 \times 10^{-6}} = 1.3 \times 10^{-3}$ mol dm^{-3}

$pH = -\log[H^+] = -\log 1.3 \times 10^{-3} = $ pH 2.88

Finding the Concentration of a Weak Acid from pH

$pH = -\log [H^+]$ $[H^+] = 10^{-pH}$

$K_a \times$ initial concentration of acid $= [H^+]^2$

Initial concentration of acid $= [H^+]^2 \div K_a$

Example
What is the concentration of ethanoic acid if the pH is 3.5?

pH 3.5 $[H^+] = 10^{-3.5} = 3.16 \times 10^{-4}$ mol dm^{-3}

Initial concentration acid =
$[H^+]^2 \div K_a = (3.16 \times 10^{-4})^2 \div 1.7 \times 10^{-5}$

Initial concentration of acid $= 5.87 \times 10^{-3}$ mol dm^{-3}

When the Two Assumptions Are Not Reliable

Unless the weak acid is very weak there will be significant dissociation at equilibrium. The initial concentration will then not equal the equilibrium concentration, e.g. as in formic acid pK_a 3.75.

When the acid solution is very dilute, the contribution of H^+ from the ionisation of water is significant. At $[H^+]$ below 10^{-6} mol dm^{-3} the calculated pH will differ from the measured pH.

Titration Curves for Acid–Base Titrations

The equivalence point for a titration is the point when the quantity of H^+ and OH^- added is equal.

pH curve for a strong acid against a strong base:

Example
25 cm^3 0.1 mol dm^{-3} HCl titrated with 0.1 mol dm^{-3} NaOH
$HCl(aq) + NaOH(aq) \longrightarrow NaCl(aq) + H_2O(l)$

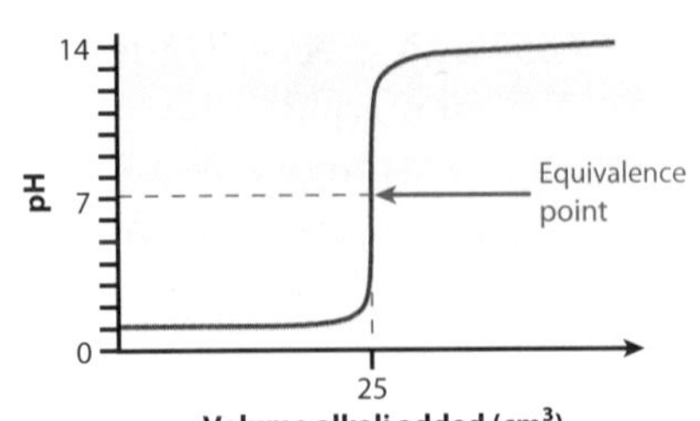

- The pH starts very low, 1–2.
- There is a vertical increase in pH from below pH 4 to above pH 10. The vertical region is centred on pH 7, the equivalence point.

Weak acid against a strong base:

Example

25 cm^3 0.1 mol dm^{-3} CH_3COOH against 0.1 mol dm^{-3} NaOH

$$CH_3COOH(aq) + NaOH(aq) \longrightarrow CH_3COONa(aq) + H_2O(l)$$

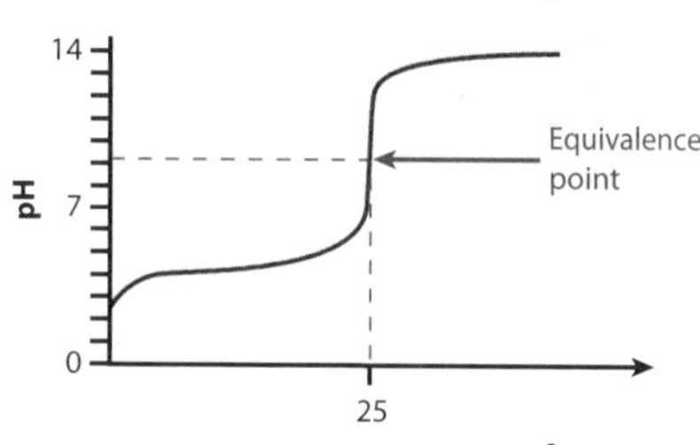

- The graph starts at a higher pH than for a strong acid (2.9 for 0.1 mol dm^{-3} ethanoic acid).
- The vertical increase in pH begins at a higher pH than for a strong acid but rises to the same pH level. The centre of the vertical region is higher than pH 7.

The salt produced by a weak acid is a conjugate base. It accepts H^+ from water leaving OH^- in solution.

$$CH_3COO^-(aq) + H_2O(l) \rightleftharpoons CH_3COOH(aq) + OH^-(aq)$$

When all the weak acid is converted to salt the resulting solution has a pH higher than 7.

Strong acid against a weak base:

Example

25 cm^3 0.1 mol dm^{-3} HCl against 0.1 mol dm^{-3} NH_3

$$HCl(aq) + NH_3(aq) \longrightarrow NH_4Cl(aq)$$

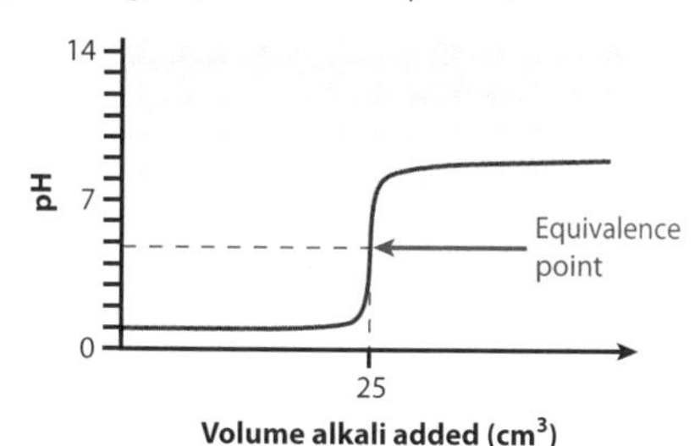

- The first part of the curve is the same as strong acid and strong base.
- The vertical section rises to pH 8–9. The centre of the vertical region is below pH 7.

The salt produced from a strong acid and a weak base is a conjugate acid. It donates H^+ to water.

$$NH_4^+(aq) + H_2O(l) \rightleftharpoons NH_3(aq) + H_3O^+(aq)$$

When all the weak base is converted to salt the resulting solution has a pH below 7.

Weak acid against a weak base:

Example

25 cm^3 0.1 mol dm^{-3} ethanoic acid against 0.1 mol dm^{-3} ammonia

$$CH_3COOH(aq) + NH_3(aq) \longrightarrow CH_3COONH_4(aq)$$

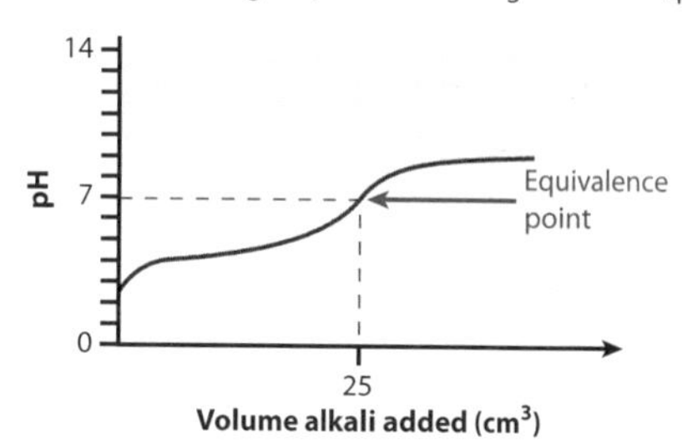

- There is no clear vertical region; the variation in pH is gradual throughout.
- At the equivalence point there is a change in direction of the curve, the point of inflection.

Indicators

Indicators are weak acids that have different colours as acid and conjugate base.

$$HI \rightleftharpoons H^+ + I^-$$

acid form	conjugate base form
colour 1	colour 2

As the [H^+] changes, the proportion of indicator in the two forms changes and the solution changes colour.

If the colour change is at a pH in the vertical section of the titration curve then it will change colour within the same drop of added acid/base and will be numerically the same as the equivalence point.

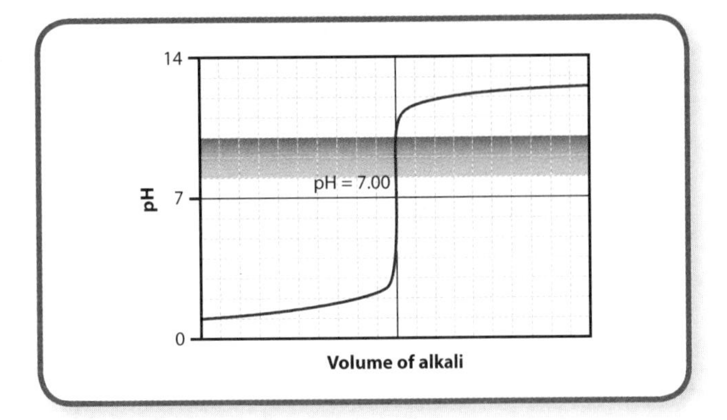

Phenolphthalein changes colour over the range pH 8.2–10. This is the vertical region for a titration of strong acid against a strong base (above) but a higher pH than the vertical region of a strong acid against a weak base (below).

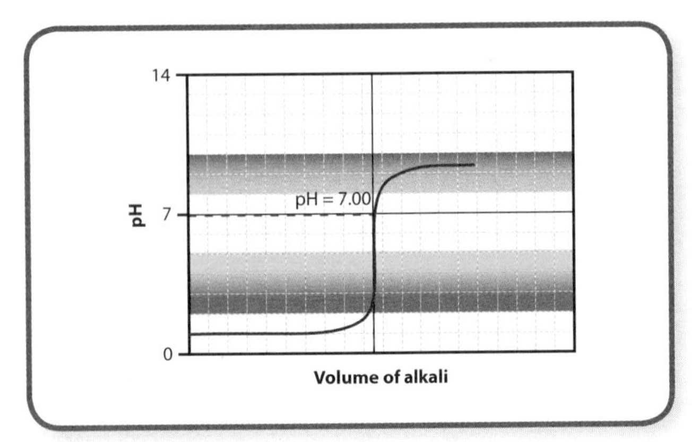

Methyl orange changes colour between pH 3.2–4.4 and is a suitable indicator.

Suitable Indicators for Acid–Base Titrations

Strong acid, strong base – many indicators phenolphthalein has a very clear colour change.

Weak acid, strong base – phenolphthalein, methyl red, thymol blue.

Strong acid, weak base – methyl orange, methyl red.

Weak acid, weak base – typically no suitable indicator; the end point must be found by plotting the pH change during the titration and finding the point of inflection of the curve.

Calculating pK_a from the Titration Curve

At the start of the titration of a weak acid with a base, only acid is present. As alkali is added, more and more salt is formed until at the equivalence point all the acid has reacted.

At the half equivalence point the concentration of acid and salt are equal, $[HA] = [A^-]$.

For a weak acid $K_a = [H^+]\frac{[A^-]}{[HA]}$

At half equivalence $\frac{[A^-]}{[HA]} = 1$

Therefore $K_a = [H^+]$ and $pK_a = pH$

The pH at the half equivalence point is equal to pK_a.

To find pK_a for a weak acid, titrate the acid with an alkali, record the pH against volume of alkali, plot the pH curve and find the pH at the half equivalence volume.

QUICK TEST

1. What is the pH of 0.200 mol dm^{-3} acid with K_a 6.3×10^{-5} mol dm^{-3}?
2. What would be the concentration of a solution of this acid with a pH of 4.00?
3. What assumptions are made when calculating the pH of a weak acid?
4. Why might this not be appropriate for an acid with a K_a of 5.0×10^{-3} mol dm^{-3}?
5. Is the pH at the equivalence point of the titration of a weak acid with a strong base higher or lower than 7?
6. Which indicator would be suitable for such a titration?

PRACTICE QUESTIONS

1. Which of the following has the highest concentration of hydrogen ions? **[1 mark]**

   A 0.1 mol dm^{-3} ethanoic acid ☐

   B 0.1 mol dm^{-3} HCl ☐

   C 0.1 mol dm^{-3} NaOH ☐

   D 0.1 mol dm^{-3} H_2SO_4 ☐

2. Which is true for the titration of a strong acid with a weak base? **[1 mark]**

   A Methyl orange is a suitable indicator. ☐

   B The equivalence point is above pH 7. ☐

   C The salt hydrolyses to release hydroxide ions. ☐

   D There is no suitable indicator. ☐

3. Vinegar contains the weak acid ethanoic acid, pK_a 4.8.

   a) Sketch a graph of volume of sodium hydroxide against pH for a titration of vinegar with sodium hydroxide solution. **[4 marks]**

   b) Explain why the pH of the solution at the equivalence point is not pH 7.0. **[2 marks]**

   c) If the pH of vinegar is 3.0 what is the concentration of ethanoic acid in vinegar? Assume that vinegar contains only ethanoic acid. **[3 marks]**

4. The indicator methyl orange is an indicator that changes colour from red to yellow. The red form of the indicator has a pK_a of 3.7.

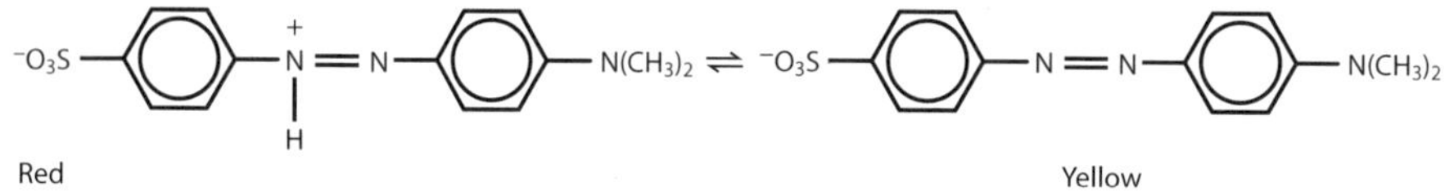

$$^{-}O_3S-C_6H_4-\overset{+}{N}(H)=N-C_6H_4-N(CH_3)_2 \rightleftharpoons {}^{-}O_3S-C_6H_4-N=N-C_6H_4-N(CH_3)_2$$

Red Yellow

   a) Why is the red form of methyl orange described as a conjugate acid? **[1 mark]**

   b) Write the expression for K_a for an indicator. Use HI to represent the indicator formula. **[1 mark]**

   c) At what pH would there be equal amounts of the red and yellow form of the indicator in solution? Explain your answer. **[2 marks]**

   d) Explain why methyl orange would not be chosen as an indicator for the titration of a weak acid against a strong base. **[2 marks]**

5. 50 cm^3 of 0.100 mol dm^{-3} of a monobasic acid was mixed with 50 cm^3 of 0.0500 mol dm^{-3} of NaOH. The [H^+] of the solution was 6.46×10^{-5} mol dm^{-3}. What is the pK_a of the acid? **[2 marks]**

Buffers

A buffer is a solution that resists change in pH when small amounts of acid or alkali are added.

Buffers contain a weak acid with its salt or a weak base with its salt.

How Buffers Work

A weak acid at equilibrium contains mainly acid with a small amount of conjugate base.

$$HA(aq) \rightleftharpoons H^+(aq) + A^-(aq)$$

If alkali is added to a weak acid it reacts with the H^+ to form water.

$$H^+(aq) + OH^-(aq) \longrightarrow H_2O(l)$$

Following Le Chatelier's principle, the position of equilibrium shifts to the right, more acid now dissociates (since K_a remains the same) and more H^+ and A^- are generated. Despite the removal of some H^+ by the alkali the concentration of H^+ remains nearly the same and so pH is almost unchanged. This is reflected in the titration curve of a weak acid: there is very little initial rise in pH as alkali is added.

If acid is added to a weak acid equilibrium then the position of equilibrium shifts to the left. The additional H^+ reacts with the conjugate base to reform the acid.

$$H^+(aq) + A^-(aq) \rightleftharpoons HA(aq)$$

The $[H^+]$ remains almost the same so the pH does not change. For a weak acid the dissociation is very small so there will be very little conjugate base present. A very small quantity of added acid will be able to use up all the conjugate base. In a buffer, more conjugate base is added in the form of the salt of the acid. The salt of the acid fully ionises, e.g.

$$CH_3COONa \longrightarrow CH_3COO^- + Na^+$$

A buffer containing CH_3COOH and CH_3COONa maintains pH when both acid and alkali are added.

$$CH_3COOH(aq) \rightleftharpoons CH_3COO^-(aq) + H^+(aq)$$

In an ideal buffer $[HA] = [A^-]$ so that the buffer can cope equally well with addition of acid and alkali.

Finding the pH of a Buffer Solution

To calculate the pH of a buffer solution two assumptions are made:

Assumption 1: The concentration of acid at equilibrium is the same as the initial concentration. There is no significant dissociation of the acid.

Assumption 2: The concentration of the conjugate base is the same as the concentration of the added salt. There is no contribution from dissociation of the acid.

The value of $K_a = [H^+] \times \frac{[\text{salt}]}{[\text{initial acid}]}$

To find pH, rearrange the equation $[H^+] = K_a \times \frac{[\text{acid}]}{[\text{salt}]}$

Example

What is the pH of a buffer made from 0.12 mol dm^{-3} ethanoic acid and 0.10 mol dm^{-3} sodium ethanoate? K_a for ethanoic acid = 1.7×10^{-5} mol dm^{-3}

$[H^+] = 1.7 \times 10^{-5} \times 0.12 \div 0.1 = 2.04 \times 10^{-5}$ mol dm^{-3}

$pH = -\log 2.04 \times 10^{-5} = 4.7$

Selecting an Acid for a Buffer

An effective buffer must have a good supply of both HA and A^- to be able to resist pH changes on addition of both acid and alkali. This means that the ratio of $HA : A^-$ should be close to 1 : 1.

When $[HA] = [A^-]$ then $pH = pK_a$

The acid chosen for a buffer of specific pH should have pK_a that is close to the desired pH.

Example

pK_a of some weak acids

Acid		pK_a
methanoic acid	$HCOOH$	3.8
benzoic acid	C_6H_5COOH	4.2
ethanoic acid	CH_3COOH	4.76
chloric acid	$HClO$	7.4
hydrocyanic acid	HCN	9.3

To make a buffer of pH 7, select chloric acid and sodium chlorate; for pH 4 use benzoic acid and sodium benzoate.

Basic Buffers

Buffers with pH greater than 7 are made from a weak base and the salt of a weak base, e.g. ammonia and ammonium chloride.

$$NH_3(aq) + H_2O(l) \rightleftharpoons NH_4^+(aq) + OH^-(aq)$$

Adding H^+ removes OH^- causing the position of equilibrium to move to the right and regenerate more OH^-.

Adding OH^- causes the position of equilibrium to move to the left, removing OH^-.

Making Buffer Solutions

A buffer solution can be made either by adding the correct mass of salt to an acid or by adding a strong base to an excess of weak acid.

Example

Making up an ethanoate buffer at pH 5.00:

$K_a = [H^+] \times \frac{[salt]}{[acid]}$ $\quad [H^+] \div K_a = \frac{[acid]}{[salt]}$ $\quad [H^+] = 10^{-pH}$

$10^{-5} = 1.7 \times 10^{-5} \times \frac{[acid]}{[salt]}$

$10^{-5} \div 1.7 \times 10^{-5} = 1 : 0.588$

Possible concentrations: ethanoic acid 0.059 mol dm^{-3} sodium ethanoate 0.1 mol dm^{-3}.

When a strong base is added to acid a salt is formed.

$$CH_3COOH + NaOH \longrightarrow CH_3COONa + H_2O$$

If the acid is in excess then the final mixture will contain both acid and salt and is a buffer.

Example

Mixing together 100 cm^3 of 0.25 mol dm^{-3} ethanoic acid with 100 cm^3 of 0.10 mol dm^{-3} sodium hydroxide:

moles = concentration × volume

moles $CH_3COOH = 0.25 \times 100 \times 10^{-3}$
$= 2.5 \times 10^{-2}$ moles

moles NaOH $= 0.1 \times 100 \times 10^{-3} = 1.0 \times 10^{-2}$ moles

The acid is in excess so all of the NaOH will react to form the salt. 1 mole NaOH forms 1 mole salt.

Moles of salt = starting moles of NaOH
$= 1.0 \times 10^{-2}$ moles

Some acid is used up in making salt. 1 mole acid forms 1 mole salt.

Moles of CH_3COONa remaining $= 2.5 \times 10^{-2} - 1.0 \times 10^{-2}$
$= 1.5 \times 10^{-2}$ moles

Total volume of solution = volume of acid + volume of base
$= 100 + 100 = 200$ cm^3

Concentration of acid = moles of acid ÷ total volume in dm^3

Concentration of acid $= 1.5 \times 10^{-2} \div 200 \times 10^{-3}$
$= 0.075$ mol dm^{-3}

Concentration of salt $= 1.0 \times 10^{-2} \div 200 \times 10^{-3}$
$= 0.050$ mol dm^{-3}

$[H^+] = K_a \times \frac{[acid]}{[salt]} = 1.7 \times 10^{-5} \times \frac{[0.075]}{[0.050]}$
$= 2.55 \times 10^{-5}$ mol dm^{-3}

$pH = -\log[H^+] = -\log 2.55 \times 10^{-5} = 4.6$

pH of the buffer formed is pH 4.6.

Diluting Buffers Has No Effect on pH

The pH of a buffer depends on the ratio of acid to salt. Adding water to a buffer dilutes both acid and salt so that the ratio remains the same and the pH does not change.

Diluting a buffer does affect how much acid or alkali it can absorb without changing pH since there is a smaller amount of both acid and salt.

Buffering in the Blood

Enzymes are very pH sensitive so it is essential that the human body regulates pH.

The healthy pH range for blood is pH 7.35–7.45. This is maintained by a buffer using the weak acid carbonic acid and its conjugate base hydrogen carbonate: $H_2CO_3(aq) \rightleftharpoons HCO_3^-(aq) + H^+(aq)$

DAY 1

Carbonic acid is formed when carbon dioxide reacts with water: $CO_2(g) + H_2O(l) \rightleftharpoons H_2CO_3(aq)$

When additional H^+ is added to blood plasma, the equilibrium moves to the right; hydrogen carbonate is converted to carbonic acid.

$$HCO_3^-(aq) + H^+(aq) \rightleftharpoons H_2CO_3(aq)$$

When additional OH^- is present it removes H^+.

$$OH^-(aq) + H^+(aq) \longrightarrow H_2O(l)$$

Carbonic acid is converted to hydrogen carbonate:

$$H_2CO_3(aq) \rightleftharpoons HCO_3^-(aq) + H^+(aq)$$

thus regenerating H^+ and restoring pH.

Many pharmaceuticals are only effective in specific pH ranges so buffers are used in making up medicines. The appearance and flavour of foods is influenced by pH, so buffers are often used in the food industry.

Using a pH Probe

A pH probe measures the concentration of H^+ in solution by comparing the test solution with a reference solution inside the probe. The difference is detected as a small voltage and converted into pH units by the meter. It must be calibrated with standard buffers before use.

Calibrating a pH Probe

- Remove the probe from the storage buffer and rinse with deionised water.
- Blot the electrode to remove surplus water and dip it into the reference buffer. The tip of the electrode must be below the surface of the solution.
- Wait for the reading to stabilise then adjust the reading to that of the buffer.
- Remove and rinse the electrode with deionised water. Blot and dip into the reference buffer a second time.
- If the reading is correct the probe is ready for use. If not, repeat until the reading is correct.

For more accurate readings two reference buffers of different pH should be used to calibrate the probe.

The pH reading will vary with temperature. All readings should be taken at the same temperature, ideally 298 K.

QUICK TEST

1. What is a buffer?
2. How does a CH_3COOH/CH_3COO^- buffer maintain its pH when OH^- is added?
3. How could you make a buffer with pH greater than 7?
4. If the pH of 0.1 mol dm^{-3} CH_3COOH/CH_3COO^- buffer is 4.0, what is the pH of a 0.01 mol dm^{-3} CH_3COOH/CH_3COO^- buffer?
5. What is the pH of a buffer made from benzoic acid and sodium benzoate where [benzoic acid] = [benzoate] and pK_a benzoic acid is 4.2?
6. Which buffering system is used to maintain the pH of blood?

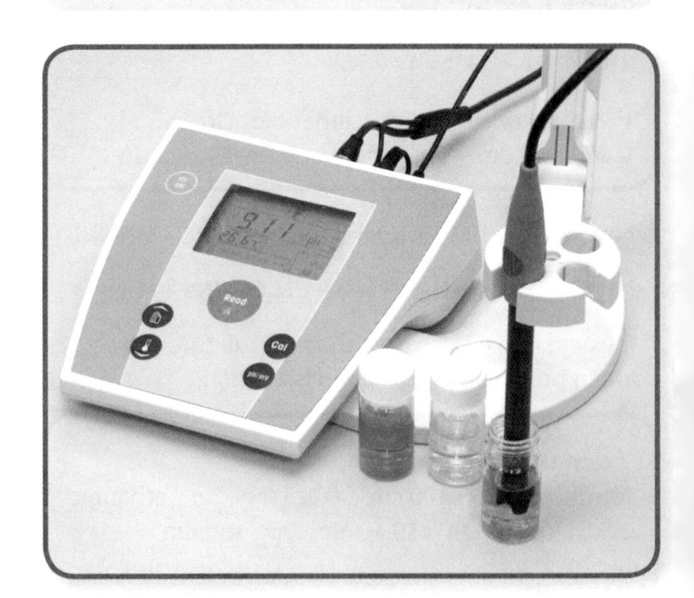

PRACTICE QUESTIONS

1. The most suitable acid for making a buffer of pH 4.2 is **[1 mark]**

		K_a/mol dm^{-3}
A	methanoic acid	1.6×10^{-4}
B	benzoic acid	6.3×10^{-5}
C	ethanoic acid	1.7×10^{-5}
D	chloric acid	3.7×10^{-8}

2. Which of the following is not true when alkali is added to a buffer made from HCOOH and $HCOO^-$? **[1 mark]**

   A The concentration of HCOOH in solution decreases.

   B The concentration of OH^- increases significantly.

   C The position of equilibrium moves towards $HCOO^-$.

   D The pH remains approximately the same.

3. A hydrogen phosphate buffer is used in many biological systems. The buffer contains the following equilibrium:

$$H_2PO_4^-(aq) \rightleftharpoons H^+(aq) + HPO_4^{2-}(aq)$$

   a) Explain how this buffer maintains the pH of a biological system. **[4 marks]**

   b) The K_a of $H_2PO_4^-$ is 6.2×10^{-8} mol dm^{-3}. Calculate the pH of a buffer containing 0.25 mol dm^{-3} $H_2PO_4^-$ and 0.1 mol dm^{-3} HPO_4^{2-}. **[2 marks]**

   c) Which would increase the pH of the buffer, increasing $[H_2PO_4^-(aq)]$ or $[HPO_4^{2-}(aq)]$? **[1 mark]**

4. Ethanoic acid buffers are suitable for a pH range of pH 3.7–5.6. K_a for ethanoic acid is 1.7×10^{-5} mol dm^{-3}.

   a) Calculate the ratio of ethanoic acid to sodium ethanoate required for a buffer of pH 4.47. **[2 marks]**

   b) What mass of NaOH should be added to 600 cm^3 of 0.1 mol dm^{-3} ethanoic acid to obtain a buffer of pH 4.47? **[4 marks]**

   c) Why is this acid not suitable for preparing a buffer of pH 7? **[2 marks]**

5. 150 cm^3 of 0.10 mol dm^{-3} methanoic acid was mixed with 50 cm^3 of 0.20 mol dm^{-3} potassium hydroxide. K_a for methanoic acid = 1.58×10^{-4} mol dm^{-3}.

   a) Calculate the pH of the solution formed. **[4 marks]**

   b) Describe the effect on pH of adding 200 cm^3 of water to this solution. **[1 mark]**

Lattice Enthalpy

Lattice enthalpy is a measure of the strength of the bonding in an ionic lattice.

Lattice enthalpy $\Delta H^{\ominus}_{LE}$ is the enthalpy change when 1 mole of solid lattice is formed from its gaseous ions.

Example
$Na^{+}(g) + Cl^{-}(g) \longrightarrow NaCl(s)$

Lattice enthalpy is always exothermic because it only involves making bonds. The more negative the value of $\Delta H^{\ominus}_{LE}$, the stronger the ionic bonding.

Born–Haber Cycles

A Born–Haber cycle for sodium chloride

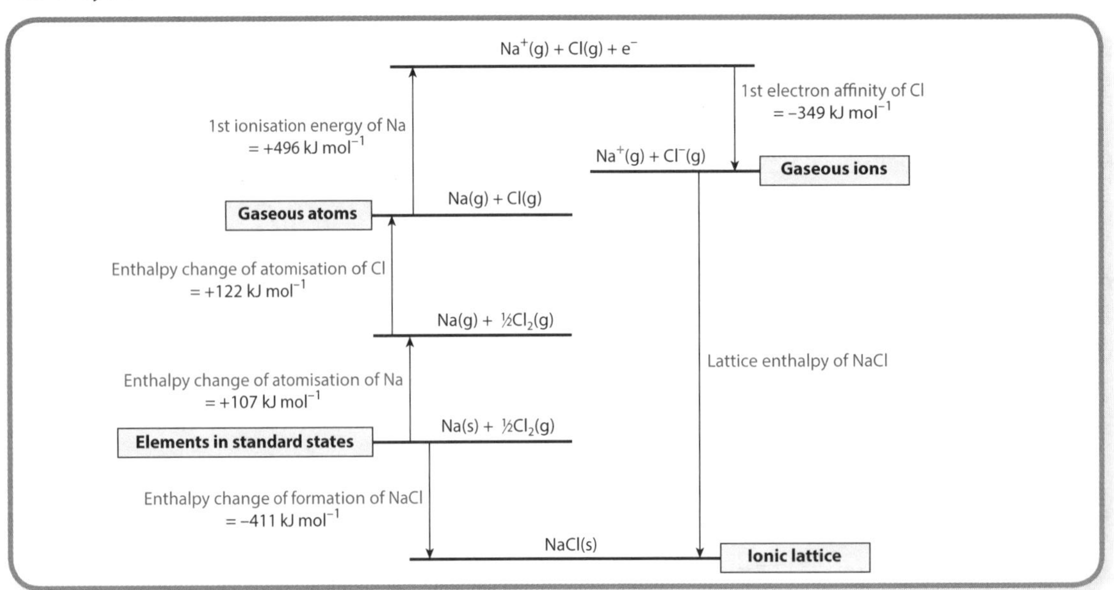

A Born–Haber cycle links together the lattice enthalpy and the standard enthalpy of formation of an ionic substance. It is drawn as an energy level diagram, which is sometimes drawn to scale (the distances between energy levels are proportional to their real values). To go from elements in their standard state to gaseous ions requires the following standard enthalpy changes:

Elements in Their Standard State to Gaseous Atoms

Atomisation enthalpy $\Delta H^{\ominus}_{atm}$ is the enthalpy change when one mole of gaseous atoms is **formed** from an element in its standard state.

$\Delta H^{\ominus}_{at}$ sodium $Na(s) \longrightarrow Na(g)$

$\Delta H^{\ominus}_{at}$ chlorine $0.5Cl_2(g) \longrightarrow Cl(g)$

Note: $\Delta H^{\ominus}_{at}$ is the formation of one mole of Cl atoms, not the atomisation of one mole of chorine molecules. It is half of the bond enthalpy Cl–Cl.

Gaseous Atoms to Gaseous Ions

The formation of positive ions is the ionisation enthalpy $\Delta H^{\ominus}_{IE}$.

First ionisation enthalpy, $\Delta H^{\ominus}_{IE}(1)$, is the formation of one mole of unipositive ions (1+) from one mole of gaseous atoms.

$\Delta H^{\ominus}_{IE}(1)$ sodium $\quad Na(g) \longrightarrow Na^{+}(g) + e^{-}$

If the ionic lattice contains a 2+ cation then both the first and second ionisation enthalpy are required.

Second ionisation enthalpy $\Delta H^{\ominus}_{IE}(2)$, is the enthalpy change when one mole of electrons is removed from one mole of gaseous 1+ ions.

$\Delta H^{\ominus}_{IE}(2)$ magnesium $\quad Mg^{+}(g) \longrightarrow Mg^{2+}(g) + e^{-}$

The formation of negative ions is the electron affinity $\Delta H^{\ominus}_{EA}$.

The first electron affinity of an element is the enthalpy change when one mole of electrons is added to one mole of gaseous atoms to form one mole of 1– ions.

$\Delta H^{\ominus}_{EA}(1)$ chlorine $\quad Cl(g) + e^- \longrightarrow Cl^-(g)$

First electron affinities are negative as energy is released when an electron is attracted to the nucleus of a neutral atom.

If the lattice contains a 2– ion then the second electron affinity is also required.

The second electron affinity is the enthalpy change when one mole of electrons is added to one mole of 1– ions to form one mole of 2– ions.

$\Delta H^{\ominus}_{EA}(2)$ oxygen $\quad O^-(g) + e^- \longrightarrow O^{2-}(g)$

Second and subsequent electron affinities are positive since a negative electron is being added to a negative ion.

Elements in Their Standard State to Ionic Lattice

Enthalpy of formation $\Delta H^{\ominus}_{f}$ is the enthalpy change when one mole of a substance is formed from its elements in their standard state.

$$Na(s) + 0.5Cl_2(g) \longrightarrow NaCl(s)$$

Drawing Born–Haber Cycles

State symbols are essential as there is an energy change when you change state. When drawing the cycle, it doesn't matter if you think about atomising and then ionising or the other way around.

Calculating Lattice Enthalpies

The lattice enthalpy can be calculated using Hess's Law: the energy change of a reaction is independent of the route taken. Lattice enthalpy energy is the difference between the gaseous ions and the solid lattice.

$$Na^+(g) + Cl^-(g) \longrightarrow NaCl(s)$$

An alternative route between these two substances can be found using the Born–Haber cycle:

LE = –(electron affinities of the anion) – (ionisation energies of the cation) – (atomisation enthalpies of the cations) – (atomisation enthalpies anions) + (enthalpy of formation)

For sodium chloride:

$$LE = -(-349) - (+496) - (+122) - (+107) + (-411) = -787 \text{ kJ mol}^{-1}$$

Born–Haber cycle for magnesium chloride

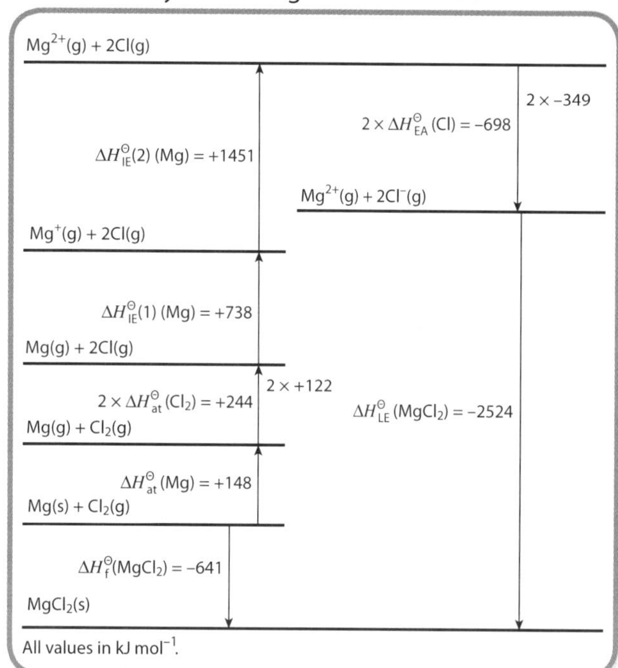

The lattice enthalpy is:

LE = –2(first electron affinity of chlorine) – (2nd atomisation enthalpy of Mg) – (1st ionisation enthalpy of Mg) – 2(atomisation enthalpy of Cl) – (atomisation enthalpy of Mg) + (enthalpy of formation of $MgCl_2$)

$$LE = -2(-349) - (+1451) - (+738) - 2(+122) - (+148) + (-641) = -2524 \text{ kJ mol}^{-1}$$

Don't forget

...there are 2 chloride ions per mole of lattice so the value for the changes in chlorine are each multiplied by 2

...Mg forms a 2+ ion so both the first and the second ionisation enthalpy of Mg are needed. The values for 1st and 2nd ionisation enthalpy are different so this is not the same as 2 × the first ionisation enthalpy.

Using Born–Haber Cycles to Calculate Other Values

Any missing value can be calculated from the Born–Haber cycle provided all the other values are known.

Example

What is the atomisation enthalpy of chlorine?

Atomisation enthalpy = forming one mole of gaseous chlorine atoms from chorine in its standard state: $0.5Cl_2(g) \longrightarrow Cl(g)$

Find an alternative route between reactants and products using the Born–Haber cycle.

2 × atomisation enthalpy of chlorine = −(atomisation enthalpy of Mg) + (enthalpy of formation of $MgCl_2$) − (lattice enthalpy) − 2(first electron affinity of chlorine) − (2nd atomisation enthalpy of Mg) − (1st ionisation enthalpy of Mg)

2 × atomisation enthalpy of chlorine = −(+148) + (−641) − (−2524) − 2(−349) − (+1451) − (+738) = 244

$\Delta H^{\ominus}_{atm}(Cl_2) = +122\ kJ\ mol^{-1}$

Note: It is important to be clear about the definitions. Remember to divide by 2 at the end of the calculation.

QUICK TEST

1. What is meant by lattice enthalpy?
2. Give the equation for the standard enthalpy of atomisation of Br_2.
3. Why is the first electron affinity negative?
4. Is the second ionisation enthalpy of Ca larger, smaller or the same as the first ionisation enthalpy?
5. Which standard enthalpy change is represented by the following equation?

   $$2Ca(s) + O_2(g) \longrightarrow 2CaO(s)$$

6. Write an equation representing the energy changes needed to calculate the first electron affinity of Cl from the Born–Haber cycle for NaCl.

PRACTICE QUESTIONS

1. The standard enthalpy of formation of NaCl is –411 kJ mol^{-1}. Which is true? **[1 mark]**

   A The lattice enthalpy for NaCl is +411 kJ mol^{-1}. ☐

   B The equation for this enthalpy change is $2Na(s) + Cl_2(g) \longrightarrow 2NaCl(s)$. ☐

   C The lattice enthalpy for NaCl is more negative than –411 kJ mol. ☐

   D The formation of sodium chloride from its elements is endothermic. ☐

2. Which calculation represents the lattice enthalpy of KF? **[1 mark]**

   A Enthalpy of atomisation of F and K + ionisation enthalpies of F and K + enthalpy of formation of KF ☐

   B Enthalpy of formation of KF – enthalpies of atomisation of F and K – ionisation enthalpies of K and F ☐

   C Enthalpy of formation of KF – enthalpies of atomisation of F and K – ionisation enthalpies of K + electron affinity of F ☐

   D Enthalpy of formation of KF – enthalpies of atomisation of F and K – ionisation enthalpies of K – electron affinity of F ☐

3. The Born–Haber cycle for CsCl is shown below.

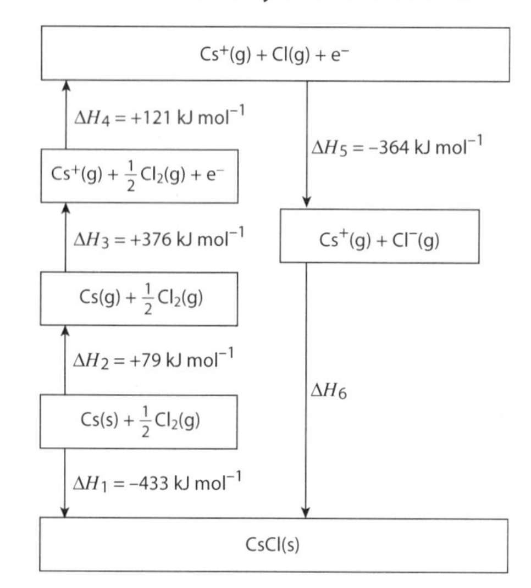

   a) Name the enthalpy changes represented by ΔH_1, ΔH_2, ΔH_3 and ΔH_4. **[4 marks]**

   b) Why does ΔH_6 have a negative value? **[1 mark]**

   c) Calculate a value for the lattice enthalpy. **[2 marks]**

4. Draw a Born–Haber cycle for calcium fluoride and label the enthalpy changes. It need not be to scale. **[6 marks]**

Theoretical Lattice Enthalpies/ Enthalpy of Solution

Theoretical Lattice Enthalpies

Oppositely charged ions attract each other and same charge ions repel each other. The strength of ionic bonding depends on the balance between this attraction and repulsion.

Theoretical lattice enthalpies can be calculated by considering the electrostatic attraction between ions of a given radius and charge.

Charge density of ions is $\frac{\text{charge on ion}}{\text{ionic radius}}$.

The higher the charge density of the ions in the lattice, the stronger the ionic bonding and the more negative the lattice enthalpy.

Charge density decreases down a group because the ionic radius increases, e.g. the lattice enthalpies of the sodium halides become less negative down the group as the ionic radius of the halide ion increases.

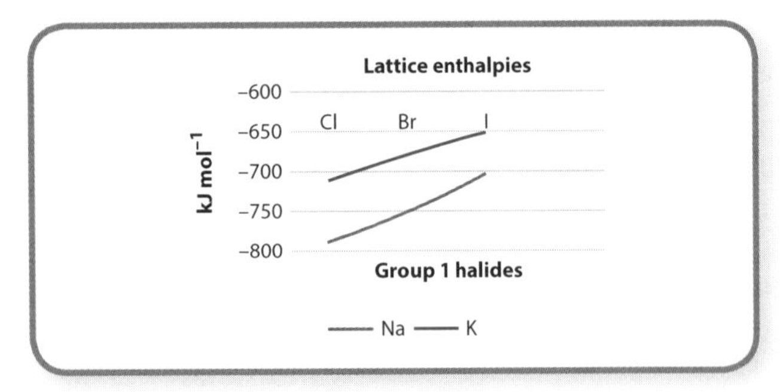

The lattice enthalpy of the potassium halides is less negative than the lattice enthalpy of the sodium halides because potassium ions have a larger radius than sodium ions.

	Lattice enthalpy kJ mol^{-1}		Lattice enthalpy kJ mol^{-1}
NaCl	−787	KCl	−711
NaBr	−751	KBr	−679
NaI	−705	KI	−651

Polarising Power and Polarisability

When cations (positive ions) and anions (negative ions) come close together, the electron cloud surrounding the anion is attracted towards the cation. This changes the spherical shape of the anion, polarising it. The larger the radius of the anion and the larger the negative charge it carries, the more easily it is polarised.

The smaller the radius and the higher the charge on the cation, the higher the charge density and the greater its polarising power.

When the ions are polarised rather than spherical the strength of the ionic bond changes and it is described as having some covalent nature. This makes the lattice enthalpy more negative than predicted from theoretical considerations. Comparing the theoretical lattice enthalpy and the lattice enthalpy as calculated by a Born–Haber cycle gives an indication of the amount of covalent nature in the ionic compound.

	Born–Haber LE kJ mol^{-1}	Theoretical LE kJ mol^{-1}	Difference kJ mol^{-1}
$MgCl_2$	−2526	−2326	200
$MgBr_2$	−2440	−2097	343
MgI_2	−2327	−1944	383

The difference between the Born–Haber value and the theoretical value of the lattice enthalpy in the magnesium halides increases down the group. This is because the size of the halide ion increases and it becomes more polarisable, giving a more covalent nature to the lattice.

Differences in Crystal Structure

The value of the lattice enthalpy also depends on the arrangement of the ions in the crystal lattice. This accounts for differences in lattice enthalpy that do not follow the expected trend.

Thermodynamic and Kinetic Stability

Most substances have a negative enthalpy change of formation and are formed by exothermic reactions. The more negative the enthalpy of formation, the more thermodynamically stable the compound.

A few substances have a positive enthalpy of formation and are known as endothermic compounds. The more positive the enthalpy of formation the less thermodynamically stable the compound.

Kinetic stability considers the activation energy that must be overcome for the reaction to occur. The rate of a reaction gives information about the kinetic stability. If the reaction described by the enthalpy of formation has a high activation energy then the reaction will happen very slowly so the reactants are kinetically stable.

Born–Haber cycles can be used to calculate the enthalpy of formation of theoretical compounds such as MgCl.

$\Delta H^{\ominus}{}_{f}MgCl_2 = -2493\ kJ\ mol^{-1}$
$\Delta H^{\ominus}{}_{f}MgCl = -770\ kJ\ mol^{-1}$

The difference in enthalpy of formation of MgCl and $MgCl_2$ reflects the difference in thermodynamic stability of the two compounds. This explains why $MgCl_2$ forms in preference to MgCl.

Enthalpy of Solution

The enthalpy of solution $\Delta H^{\ominus}{}_{(sol)}$ is the enthalpy change when one mole of a substance is completely dissolved in water under standard conditions until there is no further change in temperature on addition of more water (sometimes described as 'to infinite dilution').

Dissolving an ionic substance involves breaking ionic bonds in the lattice and forming new bonds/attractions between the ions and water.

The energy required to break up the lattice is the negative of the lattice enthalpy, also known as the lattice dissociation enthalpy.

Solid lattice ⟶ gaseous ions

The enthalpy of hydration $\Delta H^{\ominus}{}_{(hyd)}$ is the enthalpy change when one mole of gaseous ions is completely dissolved in water (to infinite dilution).

A Hess's cycle can be drawn up that connects the enthalpy of solution, lattice enthalpy and enthalpy of hydration.

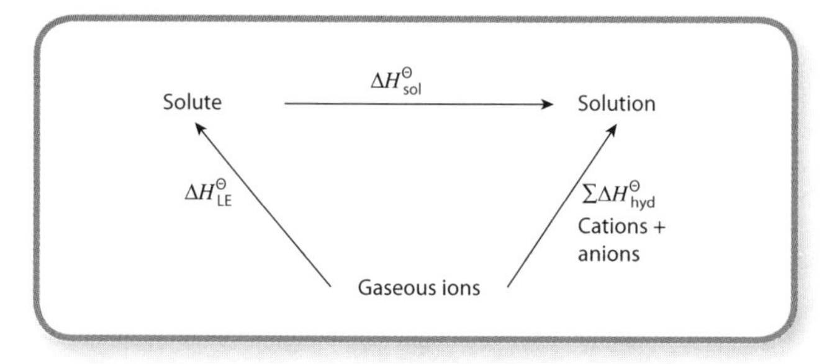

When the lattice enthalpy is more negative than the sum of the enthalpies of hydration of cation and anion then the dissolving process is exothermic. $\Delta H^{\ominus}{}_{(sol)}$ is negative.

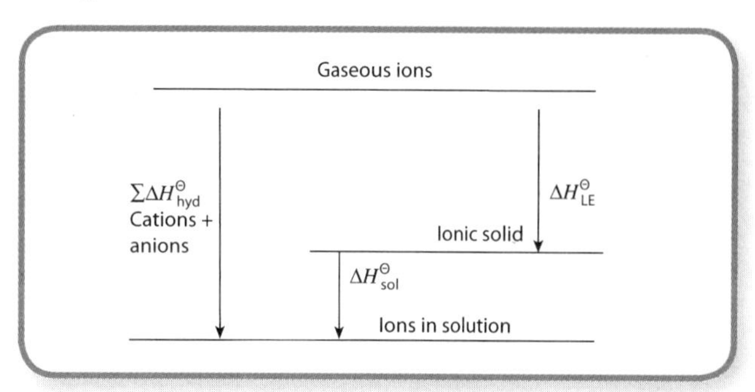

Substances with a negative enthalpy of solution are likely to be soluble.

When the lattice enthalpy is less negative than the sum of the enthalpies of solution of cation and anion then the dissolving process is endothermic. $\Delta H^{\ominus}{}_{(sol)}$ is positive.

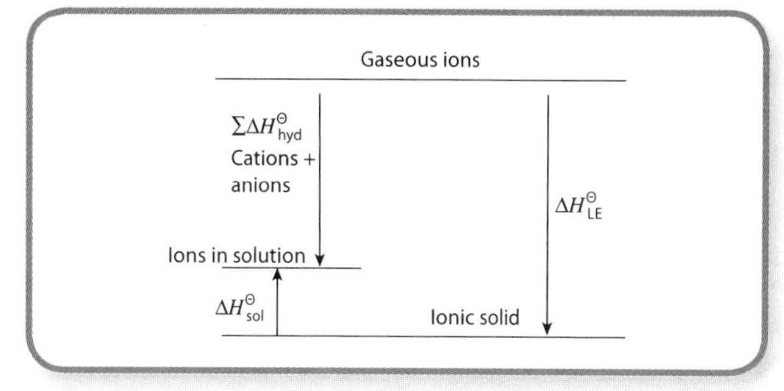

Many ionic substances have a slightly positive enthalpy of solution but still dissolve. This is because making a solution results in an increase in entropy.

Factors Affecting the Enthalpy Change of Solution

The higher the charge density of an ion, the more water molecules it is able to attract around it and the more negative the enthalpy of solution. The higher the charge density of the ions the more negative the lattice enthalpy and the stronger the ionic bonds.

Solubility of Group 2 Hydroxides

Going down Group 2, the charge density of the cations decreases as the radius increases. This causes both the enthalpy of hydration and the lattice enthalpy to become less negative. However, the rate of change in the lattice enthalpy is greater than the rate of change of the enthalpy of hydration. As a result, the enthalpy of solution decreases down the group and the hydroxides become more soluble.

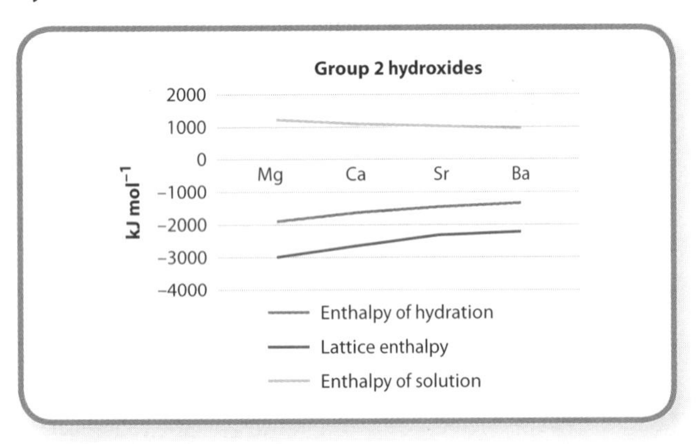

QUICK TEST

1. What does the difference between the theoretical and Born–Haber lattice enthalpy tell you?
2. What is the effect on lattice enthalpy of increasing charge density of the cation?
3. Which would have the most negative lattice enthalpy, $MgBr_2$ or MgF_2?
4. Why is there a bigger difference between theoretical and Born–Haber lattice enthalpy in CaI_2 than in CaF_2?
5. Define the term enthalpy of solution.
6. Which would have a higher enthalpy of hydration: Na^+ or Mg^{2+}?

PRACTICE QUESTIONS

1. Which of these has the most negative lattice enthalpy? **[1 mark]**

   A NaCl ☐

   B NaBr ☐

   C KCl ☐

   D KBr ☐

2. NaCl has a positive enthalpy of solution. This means that **[1 mark]**

   A it has a positive enthalpy of hydration ☐

   B the enthalpy of solution is more negative than the lattice dissociation enthalpy ☐

   C the enthalpies of hydration are more negative than the lattice enthalpy ☐

   D the enthalpies of hydration are less negative than the lattice enthalpy. ☐

3. The experimental and theoretical values for lattice enthalpy for three ionic compounds are given below.

Compound	Theoretical lattice enthalpy kJ mol^{-1}	Experimental lattice enthalpy kJ mol^{-1}
NaCl	−766	−771
AgCl	−770	−904
LiCl	−833	−853

   a) Compare the experimental lattice enthalpies for NaCl and LiCl. Explain what the difference means in terms of the ionic lattice and suggest reasons for the difference. **[3 marks]**

   b) Compare the theoretical lattice enthalpy for AgCl with its experimental value. Explain the different ways these are calculated and why they are not the same. **[3 marks]**

   c) Calculate the percentage difference in value between the experimental and theoretical values of lattice enthalpy for AgCl and NaCl. Explain what information this gives about the difference in bonding in a AgCl lattice and a NaCl lattice. **[3 marks]**

4.

Standard enthalpy of hydration Br^-/kJ mol^{-1}	−307
Standard enthalpy of hydration Ba^{2+}/kJ mol^{-1}	−1360
Lattice enthalpy $BaBr_2$/kJ mol^{-1}	−1937

   a) Draw an enthalpy cycle linking enthalpy of hydration, enthalpy of solution and lattice enthalpy for $BaBr_2$. **[2 marks]**

   b) Calculate a value for the standard enthalpy of solution of $BaBr_2$. **[2 marks]**

Entropy, S

Entropy is the degree of disorder. The second law of thermodynamics states that overall entropy always increases. ΔS is the change in entropy.

A larger number of particles means higher entropy: The more ways of arranging particles, the higher the entropy. Reactions that have a higher number of particles in the products than in the reactants cause an increase in entropy, $+\Delta S$.

Example

$CaCO_3(s) \longrightarrow CaO(s) + CO_2(g)$

The greater the freedom of movement the higher the entropy:
The entropy of a gas is higher than the entropy of a liquid, which is higher than the entropy of a solid.

State	$H_2O(g)$	$H_2O(l)$	$H_2O(s)$
entropy $J\,K^{-1}\,mol^{-1}$	189	70	45

Chemical reactions that produce a gas from solid or aqueous reactants have a $+\Delta S$.

Example

$CaCO_3(s) + 2HCl(aq) \longrightarrow CaCl_2(aq) + CO_2(g) + H_2O(l)$

A mixture has a higher entropy than a pure substance: There are more ways of arranging the particles in a mixture than in a pure substance. Dissolving an ionic solid has a positive entropy change.

Example

$NH_4NO_3(s) \longrightarrow NH_4^+(aq) + NO_3^-(aq)$

The higher the temperature the higher the entropy: The more ways of arranging quanta of energy in a substance, the higher the entropy. At higher temperatures there are more quanta of energy.

The larger the molecule the higher the entropy: There are more ways of arranging quanta of energy in a molecule with a higher M_r.

Standard molar entropy $J\,K^{-1}\,mol^{-1}$	CH_4	C_2H_6	C_3H_8
	186	230	270

Calculating $\Delta S^\ominus$ from $S^\ominus$

The entropy change $\Delta S^\ominus$ for the chemicals in a reaction can be calculated from values of $S^\ominus$, the standard molar entropy of reactants and products.

Standard molar entropy $J\,K^{-1}\,mol^{-1}$	Fe_2O_3	Fe	CO	CO_2
	87.4	27.3	197.6	213.6

$\Delta S^\ominus = \Sigma S^\ominus(\text{products}) - \Sigma S^\ominus(\text{reactants})$

$Fe_2O_3(s) + 3CO(g) \longrightarrow 2Fe(s) + 3CO_2(g)$ ΔH 24.8 kJ mol^{-1}

$\Delta S^\ominus = 3(213.6) + 2(27.3) - 87.4 - 3(197.6)$
$= +15.2\ J\,K^{-1}\,mol^{-1}$

Changing the Entropy of the Surroundings

Exothermic reactions increase the entropy of the surroundings:
When a chemical reaction occurs there is a change in entropy of the chemicals and in enthalpy. Heat energy is either released or taken in. In exothermic reactions heat energy is released. This increases the temperature and entropy of the surroundings.

Endothermic reactions decrease the entropy of the surroundings:
In endothermic reactions, heat energy is absorbed and the temperature of the surroundings decreases.

Feasibility of reactions depends on total entropy change:
The total entropy change as a result of a reaction, ΔS_{tot}, is the entropy change in the chemicals ΔS_{sys} plus the entropy change in the surroundings ΔS_{surr}

$$\Delta S_{tot} = \Delta S_{sys} + \Delta S_{surr}$$

For an event to be feasible there must be an increase in total entropy, that is (ΔS_{tot}) must be positive.

$\Delta S_{surr} = -\frac{\Delta H}{T}$ where T is the temperature in kelvin.

Exothermic Reactions

For an exothermic reaction ΔH is negative and so $-\frac{\Delta H}{T}$ is positive. ΔS_{surr} is always positive.

If ΔS_{sys} is positive, the reaction is feasible.

ΔS_{tot} = positive ΔS_{sys} + positive ΔS_{surr}

Example

$Fe_2O_3(s) + 3CO(g) \longrightarrow 2Fe(s) + 3CO_2(g)$

ΔH −24.3 kJ mol^{-1} $\Delta S = +15.2$ J K^{-1} mol^{-1}

If ΔS_{sys} is negative ΔS_{tot} is not always positive.

ΔS_{tot} = negative ΔS_{sys} + positive ΔS_{surr}

Example

$CO(g) + 2H_2(g) \rightleftharpoons CH_3OH(g)$

At 298 K $\Delta S_{sys} = -220$ J K^{-1} mol^{-1} $\Delta H = -91$ kJ mol^{-1}

$\Delta S_{surr} = -\frac{-91000}{298} = +305$ J K^{-1} mol^{-1}

Note: kJ mol^{-1} must be converted to J mol^{-1}

$\Delta S_{tot} = -220 + 305 = +85$ J K^{-1} mol^{-1}

The reaction is feasible at 298 K.

As the temperature of the reaction increases, the value of $-\frac{\Delta H}{T}$ becomes smaller.

At 413 K $\Delta S_{surr} = -\frac{-91000}{413} = +220$ J K^{-1} mol^{-1}

$\Delta S_{tot} = -220 + 220 = 0$ J K^{-1} mol^{-1}

This is the maximum temperature at which the reaction is feasible. At any temperature above 413 K

$\Delta S_{surr} < \Delta S_{sys}$ and ΔS_{tot} becomes negative.

To calculate the temperature at which the reaction becomes feasible, find the temperature at which $\Delta S_{surr} = \Delta S_{sys}$ or $\Delta S_{sys} = \frac{\Delta H}{T}$ $T = \frac{\Delta H}{\Delta S_{sys}}$

Endothermic Reactions

For an endothermic reaction ΔH is positive and so $-\frac{\Delta H}{T}$ is negative. ΔS_{surr} is always negative.

ΔS_{tot} = positive ΔS_{sys} + negative ΔS_{surr}

The value of $\frac{\Delta H}{T}$ increases as the temperature decreases until $\Delta S_{sys} = \Delta S_{surr}$. Below this temperature $\Delta S_{sys} < \Delta S_{surr}$ so ΔS_{tot} becomes negative and the reaction is not feasible.

Example

$CH_4(g) + 2H_2O(g) \rightleftharpoons CO_2(g) + 4H_2(g)$ ΔH + 164.9 kJ mol^{-1}

At 298 K $\Delta S_{sys} = -172.4$ J K^{-1} mol^{-1}; $\Delta S_{surr} = +553.4$ J K^{-1} mol^{-1}

$\Delta S_{tot} = -172.4 + 553.4 = +381$ J K^{-1} mol^{-1}

Temperature at which $\Delta S_{tot} = 0$

$T = \frac{\Delta H}{\Delta S_{sys}}$ $T = \frac{164900}{172.4} = 956.5$ K

Below 956.5 K the reaction is not feasible.

For an endothermic reaction where ΔS_{sys} is negative, ΔS_{tot} = negative ΔS_{sys} + negative ΔS_{surr}.

ΔS_{tot} can never be positive at any temperature. This reaction is not feasible (under these conditions).

Gibbs Free Energy, G

Gibbs free energy is the energy available to do useful work. For a chemical reaction to be feasible it must result in a decrease in Gibbs free energy: ΔG must be zero or negative. A reaction that results in a decrease in Gibbs free energy is described as **spontaneous**.

The change in Gibbs free energy can be calculated as $\Delta G = \Delta H - T\Delta S$ where ΔS is ΔS_{sys}.

Calculating the temperature at which a reaction is feasible:
When ΔG is less than 0 the reaction is feasible.

So when $\Delta H - T\Delta S > \Delta G$ the reaction is feasible.

When $T\Delta S > \Delta H$ the reaction is feasible.

ΔH is positive and ΔS is negative for an endothermic reaction that causes an increase in entropy.

Example

$CH_4(g) + 2H_2O(g) \rightleftharpoons CO_2(g) + 4H_2(g)$

$\Delta H = +164.9$ kJ mol^{-1}; $\Delta S = +172.4$ J K^{-1} mol^{-1}

$T \times +172.4$ must be > +164.9

$T\Delta S = \Delta H$ when $T = \frac{\Delta H}{\Delta S} = \frac{164900}{172.4} = 956.5$ K or 683.5°C

The reaction is feasible at temperatures greater than 956.5 K.

ΔH and ΔS are both negative for an exothermic reaction that causes a decrease in entropy.

Example

$$CO(g) + 2H_2(g) \rightleftharpoons CH_3OH(g)$$

$\Delta H = -91$ kJ mol^{-1}; $\Delta S = -220$ J K^{-1} mol^{-1}

$T = \frac{-91000}{-220} = 413$ K

The reaction is feasible at temperatures below 413 K.

ΔS is positive and ΔH is negative for an exothermic reaction that causes an increase in entropy.

$T\Delta S$ is always > ΔH whatever the value of T.

This reaction is feasible at all temperatures.

ΔS is negative and ΔH is positive for an endothermic reaction that causes a decrease in entropy.

$T\Delta S$ can never be > ΔH whatever the value of T.

This reaction is not feasible at any temperature (under these conditions).

A decrease in Gibbs free energy means a reaction is thermodynamically feasible but does not give information about the rate at which the reaction occurs. The rate may be so slow that no conversion of reactants into products is evident, e.g. the conversion of diamond into graphite is spontaneous/feasible but is not seen to happen within a human lifetime.

Entropy and Equilibria

For a reaction at equilibrium, ΔS_{tot} and ΔG both = 0 so there is no change in entropy or free energy.

The relationship between ΔG and the equilibrium constant K can be described as

$\Delta G = -RT \ln K$

Since R is a constant, ΔG is proportional to $-\ln K$.
At a given temperature, the more negative the value of ΔG for a reaction the higher the value of lnK.

When K = 1 lnK = 0 and $\Delta G = 0$
When K > 1 the equilibrium favours the products and ΔG is negative.
When K < 1 lnK is negative and the equilibrium favours the reactants. ΔG is positive.

QUICK TEST

1. State the sign of ΔS_{sys} for the reaction of acid with carbonate.
2. Does the melting of ice have positive or negative ΔS_{sys}?
3. Does an endothermic reaction have a positive or negative ΔS_{surr}?
4. Does the combustion of methane have positive or negative ΔG?
5. If ΔS_{sys} for the forward reaction at equilibrium is +25 J K^{-1} mol^{-1} what is the value of ΔS_{surr}?
6. $T\Delta S$ for a reaction is negative; the reaction is exothermic. Is the reaction feasible?

PRACTICE QUESTIONS

1. Which is true for the spontaneous decomposition of hydrogen peroxide? **[1 mark]**

$H_2O_2(l) \longrightarrow H_2O(l) + 0.5O_2(g)$ $\Delta H = -98$ kJ mol^{-1}

A	ΔS_{sys} is negative	ΔS_{surr} is negative	☐
B	ΔS_{sys} is negative	ΔS_{surr} is positive	☐
C	ΔS_{sys} is positive	ΔS_{surr} is negative	☐
D	ΔS_{sys} is positive	ΔS_{surr} is positive	☐

2. What is ΔS for the reaction $H_2(g) + Cl_2(g) \longrightarrow 2HCl(g)$? **[1 mark]**

	$S^{\ominus}$ J K^{-1} mol^{-1}
HCl(g)	187
$H_2(g)$	131
$Cl_2(g)$	221

A −22 J K^{-1} mol^{-1} ☐

B +22 J K^{-1} mol^{-1} ☐

C −165 J K^{-1} mol^{-1} ☐

D +165 J K^{-1} mol^{-1} ☐

3. The Haber process is used in the manufacture of ammonia.

$3H_2(g) + N_2(g) \rightleftharpoons 2NH_3(g)$ $\Delta H = -92$ kJ mol^{-1}

	$N_2(g)$	$H_2(g)$	$NH_3(g)$
$S^{\ominus}$ J K^{-1} mol^{-1}	192	131	221

a) Suggest, with a reason, whether ΔS_{sys} for the Haber process is positive or negative. **[2 marks]**

b) Use the data provided to calculate a value for ΔS_{sys}. **[1 mark]**

c) Show that this reaction is feasible at 298 K. **[3 marks]**

d) What is the maximum temperature at which this reaction is feasible? **[2 marks]**

e) The Haber process is typically carried out at temperatures above 700 K. Suggest how this is possible. **[1 mark]**

4. The thermal decomposition of zinc carbonate is endothermic.

$ZnCO_3(s) \longrightarrow ZnO(s) + CO_2(g)$ $\Delta H = +70$ kJ mol^{-1}

	$ZnCO_3(s)$	ZnO(s)	$CO_2(g)$
$S^{\ominus}$ J K^{-1} mol^{-1}	83	44	218

a) Copy and complete the table. **[3 marks]**

ΔS_{sys}	ΔH	Feasibility
positive	negative	
positive	positive	feasible at higher temperature, not feasible at low temperature
negative	negative	
negative	positive	

b) Calculate the minimum temperature to which zinc carbonate must be heated before it will decompose. **[2 marks]**

The Rate Equation

Order of Reaction With Respect To Reactants

For a reaction A + B + C ⟶ D + E

If the rate is directly proportional to the concentration of a reactant we say it is first order with respect to the species, e.g.

Rate ∝ [A] First order with respect to A

If the rate is proportional to the square of the concentration of the reactant we say it is second order with respect to the species, e.g.

Rate ∝ $[B]^2$ Second order with respect to B

If the rate is unaffected by the concentration of the reactant we say that is zero order with respect to the species, e.g.

Rate ∝ $[C]^0$ Zero order with respect to C

So, the overall rate expression would be

Rate ∝ $[A][B]^2[C]^0$

The Rate Equation

The proportional sign is changed to k, the rate constant.

Rate = $k[A][B]^2$

Any reactant that is zero order is not included in the rate equation (since $X^0 = 1$).

The rate equation cannot be predicted from the reaction equation but must be found by experiment. The coefficients in the reaction equation give no indication about the rate equation.

Example

$S_2O_8^{2-}(aq) + 2I^-(aq) \longrightarrow 2SO_4^{2-}(aq) + I_2(aq)$

Rate = $k[S_2O_8^{2-}][I^-]$

There is no relationship between $2I^-$ and the order of reaction with respect to I^-.

Overall Order of a Reaction

The overall order is found by adding together the powers in the rate equation.

Rate = $k[S_2O_8^{2-}][I^-]$ overall order = 1 + 1 = 2, a second order reaction.

The Rate Constant

When all the concentrations in the rate equation are 1 mol dm^{-3} then the rate of reaction is equal to the rate constant.

Rate = $k[S_2O_8^{2-}][I^-]$ Rate = k × 1 × 1

When the Temperature Increases k Increases

If the concentration of all reactants is kept constant and the temperature of the reaction is increased, then the rate will increase. This shows that the value of the rate constant increases with increasing temperature.

A Small E_a Means a Large k

Collision theory states that a higher rate of reaction is due to a higher frequency of successful collisions at $\geq E_a$ (activation energy for the reaction). If two reactions are carried out at the same temperature and 1 mol dm^{-3} of all reactants, the reaction with the highest E_a will be slower than the reaction with the lowest E_a.

At the same temperature, reactions with a high E_a have a smaller value of k than reactions with a low E_a.

Since the rate changes when a catalyst is added, k also changes when a catalyst is added. The concentration of a homogeneous catalyst is sometimes included in the rate equation.

Determining the Rate Equation from Data

Example 1: The table shows how the rate of reaction changed when the concentration of reactants was changed.

$2H_2(g) + 2NO(g) \longrightarrow 2H_2O(l) + N_2(g)$

	$[H_2]$/mol dm^{-3}	[NO]mol dm^{-3}	Rate/mol dm^{-3} s^{-1}
1	1.0	0.5	0.01
2	1.0	1.5	0.03
3	3.0	0.5	0.09

The order of reaction with respect to each reactant is deduced by comparing the concentration changes with the rate changes.

Compare experiment 1 with experiment 2.

$[H_2]$ does not change.
[NO] changes × 3.
The rate changes × 3.

The factor by which the concentration increases is the same as the factor by which the rate increases.

Rate ∝ [NO] **First order with respect to NO**

Compare experiment 1 with experiment 3.

[NO] remains constant.
$[H_2]$ increases × 3.
The rate increases × 9. $9 = 3^2$

Rate ∝ $[H_2]^2$ **Second order with respect to H_2**

Rate equation Rate = $k[NO][H_2]^2$

Overall the reaction is third order.

To find the order with respect to a reaction from data:

First find the order with respect to each reactant.

1. Find two experimental results where all concentrations remain constant except one.
2. Calculate the factor by which the concentration has changed (concentration in the second experiment ÷ concentration in the first experiment).
3. Calculate the factor by which the rate has changed (rate in the second experiment ÷ rate in the first experiment).
4. Compare the change of rate factor with the change of concentration change.

   If there has been no change in rate, the order with respect to that reactant = zero order.

   If the two factors are the same = first order.

   If the rate change is (concentration change)^2 = second order.

Example 2: Rate results for the reaction

$$CH_3I + OH^- \longrightarrow CH_3OH + I^-$$

	$[CH_3I]$/ mol dm^{-3}	$[OH^-]$/ mol dm^{-3}	Rate/ mol dm^{-3}s^{-1}
1	0.1	0.1	1×10^{-5}
2	0.2	0.1	2×10^{-5}
3	0.2	0.4	2×10^{-5}

Finding the rate with respect to CH_3I:

1. Experiments 1 and 2, only $[CH_3I]$ changes.
2. Factor by which the $[CH_3I]$ has changed 0.2 ÷ 0.1 = × 2
3. Factor by which the rate has changed 2 ÷ 1 = × 2
4. The rate change factor is the same as the concentration change factor 2 : 2 Rate ∝ $[CH_3I]$
   The reaction is first order with respect to CH_3I.

Finding the rate with respect to OH^-:

1. Experiment 2 and 3, only $[OH^-]$ changes.
2. Factor by which $[OH^-]$ has changed 0.4 × 0.1 = × 4
3. The rate does not change.
   The reaction is zero order with respect to OH^-

   Rate ∝ $[OH]^0$

Rate equation Rate = $k[CH_3I]$

Finding Rate Equations When More Than One Concentration Changes

Sometimes the data provided shows results where the concentration of more than one reactant has changed, e.g.

	[A]/ mol dm^{-3}	[B]/ mol dm^{-3}	[C]/ mol dm^{-3}	Rate/mol dm^{-3}s^{-1}
1	0.40	0.20	0.80	1.60
2	0.40	0.80	0.80	6.40
3	0.80	1.60	0.80	51.2
4	0.80	0.80	0.40	25.6

[B] changes with each experiment. There is no experiment where only [A] or [C] changes.

1. Between experiments 1 and 2:

   [A] and [C] do not change; [B] increases × 4. Rate increases × 4.

   The reaction is first order with respect to B.

2. Between experiments 2 and 3:

   [C] does not change; [A] doubles; [B] doubles.

   Rate increases × 8.

   To calculate the rate change that was caused by [A] alone we must take into account the rate increase caused by doubling [B].

   From 1, we know that rate ∝ [B].

   Effect of doubling [B] alone would be to double the rate from 6.4 to 12.8.

   The actual rate change was from 6.4 to 51.2.

   The change between 12.8 and 51.2 was caused by doubling [A].

   If [A] doubled alone the rate would change by a factor of 51.2 ÷ 12.8 = 4

   Rate ∝ $[A]^2$.

   The reaction is second order with respect to A.

3. Between experiments 2 and 4:

   [B] does not change; [A] doubles; [C] halves.

   We know that rate ∝ $[A]^2$.

   The effect of doubling [A] alone would be to increase the rate × 4 from 6.4 to 25.6.

   The actual rate change is also from 6.4 to 25.6.

   Changing [C] has had no effect on rate.

   The reaction is zero order with respect to C.

   Rate equation Rate = $k[A]^2[B]$

Summary

1. Find two experimental results where only one reactant concentration changes [*x*] and determine the order with respect to *x*.
2. Find two experiments where only [*x*] and one other reactant [*y*] changes.
3. Use the order with respect to *x* to calculate the expected rate if only [*x*] had changed.
4. Compare this expected rate with the actual rate in the data to find the rate change caused by changing [*y*].

Calculating the Rate Constant

Once the rate equation is known it can be rearranged to find the value of the rate constant, e.g.

Rate = $k[S_2O_8^{2-}][I^-]$ K = rate ÷ $[S_2O_8^{2-}][I^-]$

The values of concentration and rate from any reaction can be used in the equation to find the value of k at that temperature.

Units of k

The units of k vary according to the order of the reaction. They can be calculated from the equation k = rate(mol dm^{-3} s^{-1}) ÷ $[S_2O_8^{2-}]$(mol dm^{-3})$[I^-]$(mol dm^{-3})

Units of k = mol dm^{-3} s^{-1} ÷ mol dm^{-3} × mol dm^{-3}
= mol^{-1} dm^3 s^{-1}

QUICK TEST

1. Write the rate equation for a reaction where the order of reaction with respect to A is 1 and with respect to B is 2.
2. What is the overall order of this reaction?
3. If the rate of a reaction changes from 1.3 mol dm^{-3} s^{-1} to 5.2 mol dm^{-3} s^{-1} when the concentration of a reactant doubles, what is the order of reaction with respect to that reactant?
4. If the order of reaction with respect to A is first order and the order with respect to B is second order, by how much does the rate increase when the concentration of both A and B are doubled?

PRACTICE QUESTIONS

1. The rate equation for a reaction is Rate = $k[Br^-][BrO_3^-][H^+]^2$. The units of k are **[1 mark]**

A $mol^{-3}\,dm^{9}\,s^{-1}$ ☐

B $mol^{3}\,dm^{-9}\,s^{-1}$ ☐

C $mol^{-3}\,dm^{-9}\,s^{-1}$ ☐

D $mol^{3}\,dm^{9}\,s^{-1}$ ☐

2. The value of k for reaction A is 2.5 s^{-1} and for reaction B is 4.0 s^{-1}. Which pair of statements is true? **[1 mark]**

	The value of E_a	The rate when all concentrations are 1 mol dm^{-3}	
A	A is higher than B	A is higher than B	☐
B	A is lower than B	A is higher than B	☐
C	A is higher than B	A is lower than B	☐
D	A is lower than B	A is lower than B	☐

3. The initial rate of reaction between A and B was measured in a series of experiments and the results are shown below.

Experiment	[A]/mol dm^{-3}	[B]/mol dm^{-3}	Rate/mol $dm^{-3}s^{-1}$
1	0.20	0.20	1.2×10^{-3}
2	0.20	0.40	2.4×10^{-3}
3	0.10	0.40	0.60×10^{-3}
4	0.10	0.20	

a) Find the order of reaction with respect to A and B. Show your working. **[4 marks]**

b) Complete the missing rate value from experiment 4. **[1 mark]**

c) Write the rate equation for this reaction. **[1 mark]**

d) Calculate the value of the rate constant and state the units. **[2 marks]**

4. Rate = $k[A]^2[B]$ is the rate equation for the reaction $A + B + 2C \longrightarrow D + E$.

In an experiment the rate of reaction was found to be 4.0×10^{-4} mol dm^{-3} s^{-1}.

a) How will the value of k change if the temperature is increased? **[1 mark]**

b) Calculate the rate when the concentration of C doubles. **[1 mark]**

c) Calculate the rate when the concentration of A doubles. **[1 mark]**

d) Calculate the rate when the concentration of both A and B double. **[1 mark]**

Experiments to Find Orders of Reaction

Following the Rate of a Reaction

The rate of a reaction can be followed by monitoring the appearance of products or the disappearance of reactants with time. The rate of solution-based reactions is usually mol $dm^{-3}\ s^{-1}$.

Choosing a Method

- Coloured reactants or products: use a colorimeter.
- Gases: collect and measure volume in a gas syringe or using an inverted burette over water, or monitor mass loss if gases are lost to the atmosphere.
- H^+ or OH^-: use a pH meter.

Some reactions do not have a reactant or product that can easily be measured continuously. These can be monitored by titration.

Samples are removed from the reaction mixture at timed intervals. The reaction in the sample is stopped at the time it is removed by quenching. The quantity of one of the reactants or products in the sample can then be measured by titration.

The method of quenching depends on the reaction. It involves removing one reactant by adding something that it reacts with instantly, e.g. removing H^+ as a reactant (or catalyst) by adding alkali; removing iodine by adding thiosulfate.

When there is no reactant that can be easily removed, the reaction can be quenched by plunging the reaction vessel into ice-water, which rapidly reduces the rate.

Finding the Rate Equation Using the Initial Rates Method

A separate experiment must be carried out to find the rate with respect to each reactant. For example:

$$A + B \longrightarrow C + D$$

To find the order with respect to A:
Choose a minimum of five different concentrations of A. For each concentration, add an excess of B and follow the course of the reaction with time. Plot the results on a graph of concentration against time.

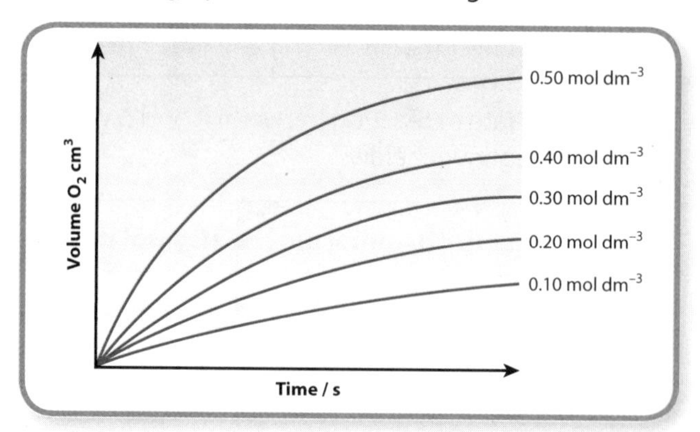

Find the initial rate of each reaction by taking a tangent to the curve at $t = 0$ and calculating the gradient.

Using an excess of B means that the concentration of B does not decrease significantly during the reaction and the rate is controlled by the concentration of A.

Plot a graph of rate versus concentration of A.

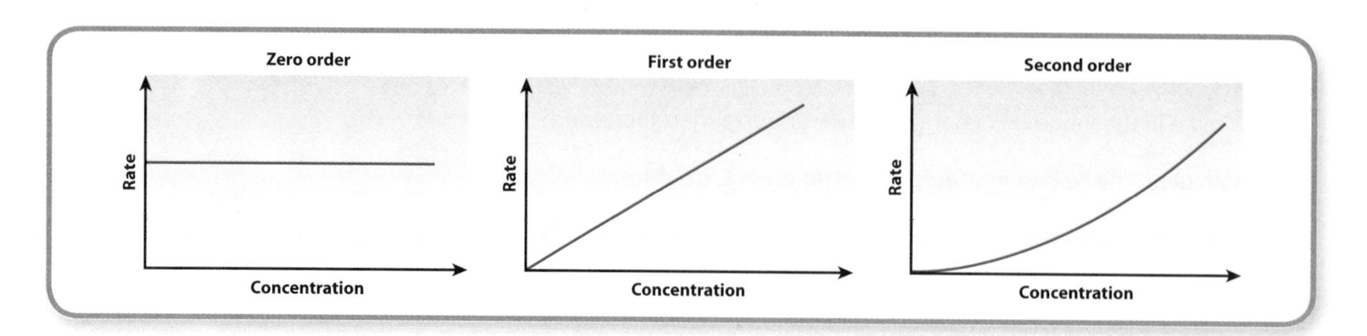

The shape of the graph gives the order of the reaction with respect to A.

Repeat the experiments using different concentrations of B and keeping A in excess to find the order with respect to B.

All experiments must be carried out at the same temperature.

A second order relationship can be confirmed by plotting a graph of rate against [concentration]^2, which gives a straight line.

Finding the Rate Equation Using the Half-life

The half-life of a reactant is the time taken for the concentration of that reactant to drop to half its value.

For a first order relationship the half-life ($t_{\frac{1}{2}}$) is constant throughout the reaction.

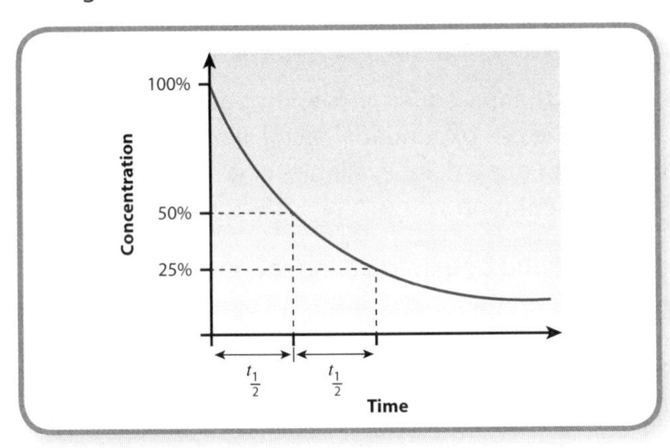

The concentration vs time graph for a zero order relationship gives a straight line. $t_{\frac{1}{2}}$ gets smaller as the reaction proceeds.

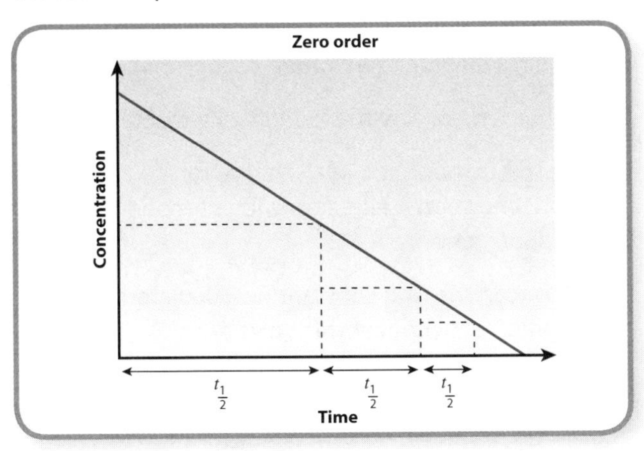

A graph of concentration vs time for a second order relationship gives a curve with an increasing half-life.

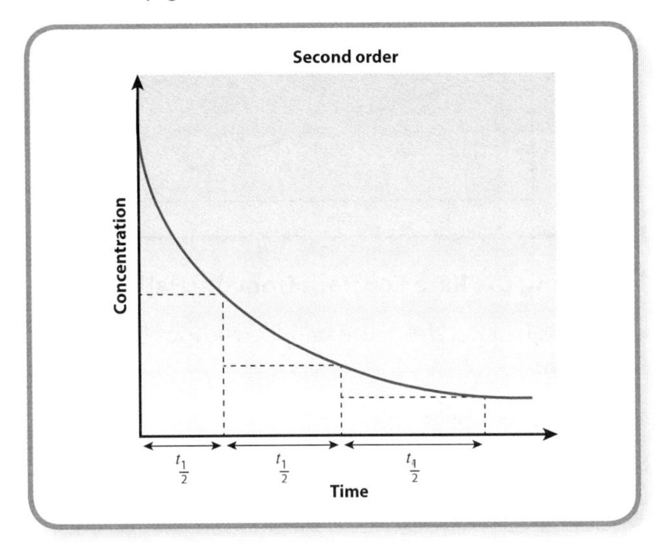

Calculating the Rate Constant from Graphs

There are two ways of calculating k for a first order reaction.

Finding k from the Rate vs Concentration Graph

For a first order relationship, rate = k[A], k = rate/[A]

k is the gradient of a graph of rate vs [A] $y \div x$.

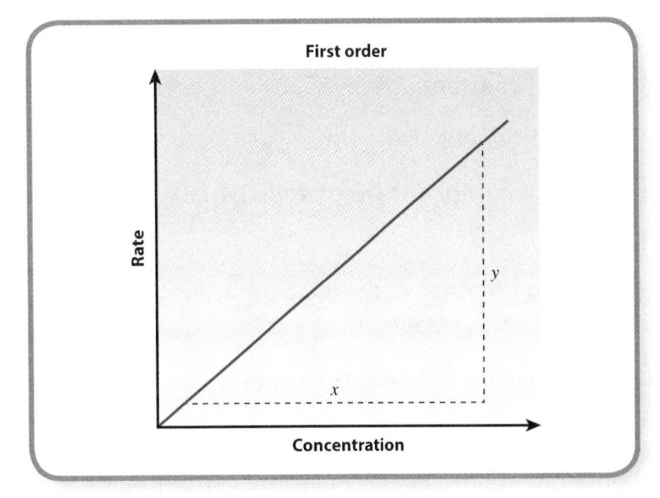

Finding k from the ln Concentration vs Time Graph

For a first order reaction k = ln concentration/time

This means that the rate constant is the gradient of the graph of ln concentration vs time.

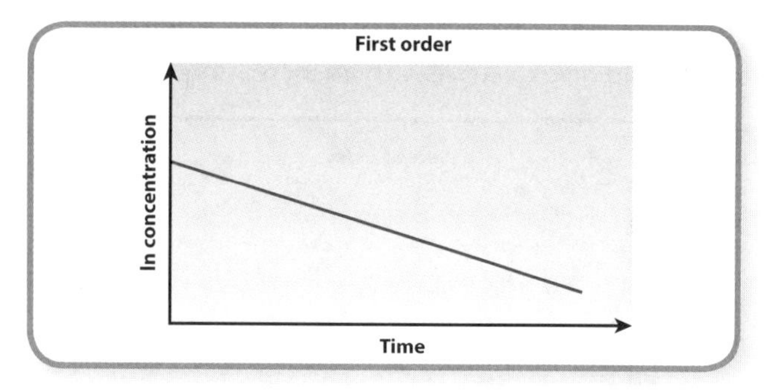

Calculating the Rate Constant from the Half-life

k can be calculated from the half-life without the need of a graph.

For a first order reaction k = ln 2 ÷ $t_{\frac{1}{2}}$
(For A-level you do not need to know why.)

ln2 = 0.693
Once the half-life is known k can be calculated.
k = 0.693 ÷ half-life

Finding Activation Energy Using the Arrhenius Equation

The equation $k = A\,e^{\frac{-E_a}{RT}}$ links the rate constant to the activation energy.

Taking ln of both sides and rearranging gives:

$\ln k = \frac{-E_a}{RT} + \ln A$

or $\ln k = \frac{-E_a}{R} \times (\frac{1}{T}) + \ln A$

R is the gas constant.

T is the temperature in Kelvin.

If a graph of lnk against $\frac{1}{T}$ is plotted then the gradient of the slope is $\frac{-E_a}{R}$

Multiplying the value of the gradient by $-R$ gives the value of E_a for the reaction.

Using Experimental Results to Find E_a

Find the rate equation for the reaction by experiment.

Choose one set of values for reactant concentration and measure the rate of this reaction at different temperatures.

Use the rate equation to calculate the value of k at each temperature.

Calculate lnk and $\frac{1}{T}$ for each experiment and plot against each other on a graph.

Find the gradient by dividing lnk by $\frac{1}{T}$. This is $\frac{-E_a}{R}$

Multiply the value by R, which is 8.314. This is $-E_a$ in joules. To find E_a in kJ, divide by 1000.

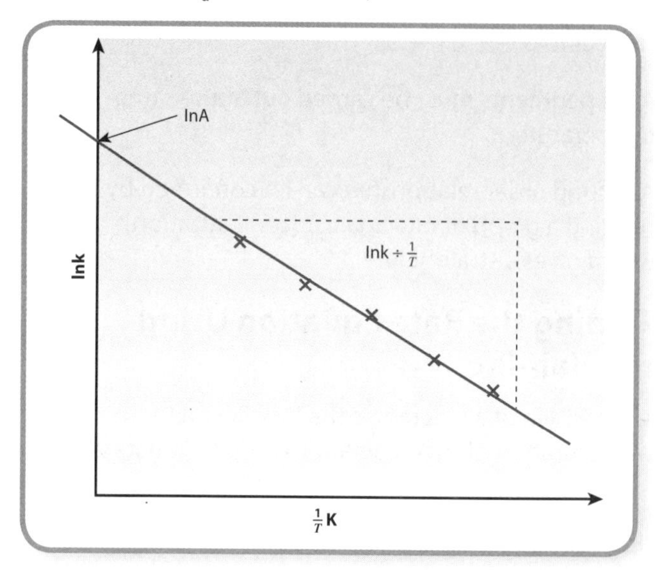

Determination of A, the Pre-exponential Factor

A in the Arrhenius equation is known as the frequency factor or the pre-exponential factor. It represents the number of particles available for collision at the correct orientation.

It can be found by extrapolating the line for lnk vs $\frac{1}{T}$ until it crosses the y-axis. This is lnA or e^A.

QUICK TEST

1. How could you follow the rate of the reaction between calcium carbonate and hydrochloric acid?
2. Sketch the shape of the graph for concentration against rate for a zero order reaction.
3. What is meant by the half-life of a reactant?
4. Describe the shape of the curve for concentration against time for a second order reaction.
5. How can the rate constant be calculated from a graph of ln concentration vs time?
6. What experimental data is needed to calculate the activation energy of a reaction from its rate?

PRACTICE QUESTIONS

1. Which of the following methods could **not** be used to follow the rate of the reaction between $MgCO_3$ and HCl? **[1 mark]**

   A Measuring the volume of gas produced ☐

   B Using a pH probe ☐

   C Measuring the change in absorbance ☐

   D Measuring the loss in mass ☐

2. Which is true of a first order reaction? **[1 mark]**

   A The concentration against time graph has a constant half-life. ☐

   B The concentration against time graph is a straight line. ☐

   C The concentration against time graph has an increasing half-life. ☐

   D The concentration against time graph has a decreasing half-life. ☐

3. The equation for the iodination of propanone is shown below.

$$CH_3COCH_3(aq) + I_2(aq) \longrightarrow CH_3COCH_2I(aq) + HI(aq)$$

   a) What property of this reaction means that the rate can be followed using a colorimeter? **[1 mark]**

   b) Describe the experiments that a student has to carry out to find the rate of reaction with respect to iodine. **[4 marks]**

   When the student plotted a graph of concentration of iodine against time she found that she got a horizontal straight line.

   c) What does this demonstrate about the order of the reaction with respect to iodine? **[1 mark]**

   Another student explored the effect of changing the concentration of propanone on the rate. He found that the reaction was first order with respect to propanone.

   d) Sketch a graph of the concentration of propanone against rate. **[1 mark]**

   A third student found that changing the concentration of the acid catalyst also affected the rate of reaction. The shape of the concentration against rate graph for acid was a curve. The student thought that the reaction was probably second order with respect to hydrogen ions.

   e) Suggest a way that this proposal could be confirmed. **[2 marks]**

   f) Write the rate equation for the reaction and state the overall order of the reaction. **[2 marks]**

4. A student measured the rate of a reaction at different temperatures, keeping the concentrations the same each time. The results were used to calculate the value of the rate constant at each temperature. A plot of lnk against $\frac{1}{T}$ for the reaction gave a line with a gradient of $-5200\,K$.

   The Arrhenius equation is $\ln k = -\frac{E_a}{RT} + \ln A$
   $R = 8.31\ J\,K^{-1}\,mol^{-1}$

   a) Calculate the value of E_a for this reaction. Include appropriate units in your answer. **[2 marks]**

   b) Explain how the value for the pre-exponential factor (frequency factor) could be found from this data. **[1 mark]**

Reaction Mechanisms

A chemical reaction involves breaking and making bonds. Bonds may be broken and made simultaneously or the reaction may take place in a series of steps. The description of the steps of a reaction is called the reaction mechanism.

In a multistep reaction the overall rate of reaction is limited by the rate of the slowest step. This is known as the rate-determining step or rate-limiting step.

The Rate-Determining Step and the Rate Equation

The rate equation shows which particles participate in the rate-determining step.*

Tertiary haloalkanes react with hydroxide ions to form alcohols. The reaction mechanism has two steps.

Overall: $(CH_3)_3CBr + OH^- \longrightarrow (CH_3)_3COH + Br^-$

Rate equation: Rate = $k[(CH_3)_3Br]$

Step 1

$H_3C{-}C(CH_3)_2{-}Br \longrightarrow (CH_3)_2C^+{-}CH_3 + Br^-$

Step 2

$(CH_3)_2C^+{-}CH_3 + {:}OH^- \longrightarrow H_3C{-}C(CH_3)_2{-}OH$

The first step is slow; it is the rate-determining step.

The first step makes a **carbocation intermediate**. Species made in one step of a reaction and then used up in another step are called intermediates.

The carbocation is converted into product by the second step, which is fast.

Increasing the concentration of the haloalkane increases the rate of the rate-determining step and so the overall rate of reaction.

*Only true where the rate-determining step is the first step of the mechanism.

Hydroxide ions are not needed for the first step so increasing the concentration of hydroxide ions will have no effect on the overall reaction rate. The reaction is zero order with respect to hydroxide ions. Hydroxide ions do not appear in the rate equation.

A nucleophilic substitution reaction where only one species is involved in the rate-determining step is described as an **S_N1** reaction.

Primary haloalkanes also react with hydroxide ions to form alcohols.

Overall: $CH_3Br + OH^- \longrightarrow CH_3OH + Br^-$

Rate equation: Rate = $k[CH_3Br][OH^-]$

This time the reaction is also first order with respect to hydroxide ions. The reaction mechanism has one step.

$H{-}CH_2{-}Br + {:}OH^- \longrightarrow H{-}CH_2{-}OH + Br^-$

The reaction is considered to go through a theoretical **transition state** where the C—Br bond is not fully broken and the C—OH bond is not fully formed.

$OH^- + H{-}CH_2{-}Br \rightarrow [HO\cdots CH_3 \cdots Br]^- \rightarrow H{-}CH_2{-}OH + Br^-$

Both CH_3Br and OH^- are needed for the first and only step. Increasing the concentration of either will increase the rate of the reaction and this is reflected in the rate equation.

Nucleophilic substitution reactions that involve two species in the rate-determining step are described as **S_N2** reactions.

Deducing a Possible Mechanism from the Rate Equation

Once the rate equation has been found, it is possible to propose a mechanism that is consistent with the rate equation. It is not possible to know if the proposed mechanism is correct but the rate equation can rule out some mechanisms.

Points to consider when proposing a mechanism:

- If the rate equation does not include all the reactants then the reaction is multistep.
- If there are more than two reactants the reaction will be multistep. (There is a very low probability that three particles would collide to cause a reaction.)
- If the reaction is second order with respect to one of the reactants then the rate-determining step involves two particles of this reactant.*
- The sum of the reactants of each step = reactants in the reaction equation. The sum of the products of each step = the products in the reaction equation.

The examples below assume that the first step is the rate-determining step, and there are only two steps. These are the easiest mechanisms to understand and the most likely to be encountered at A-level.

Example 1

$2A + B \longrightarrow C + D$ **Rate = k[A][B]**

First write in all the known information about the reaction.

The species that appear in the rate equation are the reactants of the first step.

Step 1 $A + B \longrightarrow$

Any reactants that do not appear in the rate equation are reactants in the second step.

Step 1 $A + B \longrightarrow$

Step 2 $A + \quad \longrightarrow$

Products of the first step often include one of the products of the overall reaction plus an intermediate.

Step 1 $A + B \longrightarrow C + I$

Step 2 $A + \quad \longrightarrow$

The intermediate made in the first step becomes a reactant in the second step. Any products in the reaction equation that do not appear as products of the first step must be products of the second step.

Step 1 $A + B \longrightarrow C + I$

Step 2 $A + I \longrightarrow D$

The individual steps add up to the reaction equation.

Step 1 $A + B \longrightarrow C + \cancel{I}$

Step 2 $A + \cancel{I} \longrightarrow D$

Overall: $2A + B \longrightarrow C + D$

*Only true where the rate-determining step is the first step of the mechanism.

Example 2

$2A + B \longrightarrow C + D$ **Rate = $k[A]^2$**

Two particles of A must take part in the rate-determining step. B does not participate in the rate-determining step.

Step 1 $2A \longrightarrow C + I$ $\qquad 2A \longrightarrow C + \cancel{I}$

Step 2 $B + I \longrightarrow D$ $\qquad B + \cancel{I} \longrightarrow D$

Overall: $2A + B \longrightarrow C + D$

Example 3

$A + B + C \longrightarrow D + E$ **Rate = $k[A]^2[B]$**

C does not participate in the rate-determining step. Two particles of A are required in the rate-determining step.

Step 1 $2A + B \longrightarrow AB + D$ $\qquad \cancel{2}A + B \longrightarrow \cancel{AB} + D$

Step 2 $AB + C \longrightarrow E + A$ $\qquad \cancel{AB} + C \longrightarrow E + \cancel{A}$

Overall: $A + B + C \longrightarrow D + E$

DAY 3

Two particles of A are needed for the first step but there is only one particle in the reaction equation. This means that one particle of A must be regenerated in order to balance the reaction equation.

If the rate-determining step is not the first step, all reactants that are in the steps preceding the rate-determining step must appear in the rate equation.

The Iodination of Propanone in Acidic Conditions

Overall: $CH_3COCH_3 + I_2 \longrightarrow CH_3COCH_2I + HI$

Rate equation: Rate = $k[CH_3COCH_3][H^+]$

The rate equation shows that I_2 is not involved in the rate-determining step.

Possible mechanism:

1. $(CH_3)_2C{=}O + H^+ \longrightarrow (CH_3)_2COH^+$
2. $(CH_3)_2COH^+ \longrightarrow CH_3C(OH)CH_2 + H^+$
3. $CH_3C(OH)CH_2 + I_2 \longrightarrow CH_3COCH_2I + HI$

The rate of the reaction can be followed by measuring the rate of disappearance of iodine using a colorimeter or by titration.

Titration Method

The reaction is quenched by adding sodium carbonate solution to a sample removed from the reaction mixture at timed intervals. This removes H^+. The samples are titrated with sodium thiosulfate to determine the concentration of iodine remaining.

$$I_2 + 2S_2O_3^{2-} \longrightarrow 2I^- + S_4O_6^{2-}$$

A starch indicator is added towards the end of the titration, which reacts with the remaining iodine to give a blue-black colour. The blue-black colour changes to colourless at the end point.

Harcourt–Essen Iodine Clock

A good approximation of the initial rate of a reaction is given using a clock reaction, e.g.

$$2I^- + S_2O_8^{2-} \longrightarrow I_2 + 2SO_4^{2-} \qquad \text{Reaction 1}$$

The time taken for the reaction to produce a set quantity of I_2 is measured by adding a known quantity of sodium thiosulfate to the reaction mixture. This reacts with the iodine released by the reaction.

$$I_2 + 2S_2O_3^{2-} \longrightarrow 2I^- + S_4O_6^{2-} \qquad \text{Reaction 2}$$

The quantity of thiosulfate added is calculated to be no more than is needed to react with 10% of the total iodine produced by the reaction mixture. Starch indicator is also added.

As the reaction proceeds, the iodine produced by reaction 1 is immediately used up by reaction 2. Once all the added thiosulfate has reacted, reaction 2 stops. The next portion of iodine produced by reaction 1 reacts with the starch and a blue-black colour appears.

The rate of reaction $\propto$ 1/time taken for the colour to appear. The rate of reaction in mol dm^{-3} s^{-1} = 0.5 × moles S_2O^{2-} ÷ time for the colour to change.

QUICK TEST

1. What is meant by the rate-determining step?
2. How many particles react together in the rate-determining step of an S_N2 reaction?
3. Which has an intermediate, an S_N1 or an S_N2 reaction?
4. Draw the transition state for an S_N2 reaction for the conversion of ethanol to chloroethane using HCl.
5. For the reaction $X + 2Y \longrightarrow Z + W$, how many particles of Y participate in the rate-determining step?
6. How is the rate of reaction determined in the iodine clock reaction?

PRACTICE QUESTIONS

1. $2NO_2Cl$ breaks down to $2NO_2 + Cl_2$. A proposed mechanism for this reaction is

$NO_2Cl \longrightarrow NO_2 + Cl$ slow

$NO_2Cl + Cl \longrightarrow NO_2 + Cl_2$ fast

A rate equation that is consistent with this mechanism is **[1 mark]**

A Rate = $k[NO_2Cl]^2$ ☐

B Rate = $k[NO_2Cl]$ ☐

C Rate = $k[NO_2Cl][Cl]$ ☐

D Rate = $k[NO_2][Cl]$ ☐

2. The reaction of a bromoalkane with sodium hydroxide proceeds via an S_N2 mechanism. Which of these statements is correct about this reaction? **[1 mark]**

A The reaction involves the formation of a carbocation. ☐

B The rate equation is rate = k[bromoalkane] ☐

C The mechanism is multistep. ☐

D The reaction has only one step. ☐

3. The iodination of propanone proceeds through a number of steps. The rate equation for the reaction is rate = $k[CH_3COCH_3][H^+]$. One suggested mechanism is shown here:

Step 1 $(CH_3)_2C{=}O + H^+ \longrightarrow (CH_3)_2COH^+$

Step 2 $(CH_3)_2COH^+ \longrightarrow CH_3C(OH)CH_2 + H^+$

Step 3 $CH_3C(OH)CH_2 + I_2 \longrightarrow CH_3COCH_2I + HI$

a) Write the overall equation for this reaction. **[1 mark]**

b) Explain the role of H^+ in this reaction. **[2 marks]**

c) What is meant by the rate-limiting step of a mechanism? **[1 mark]**

d) Use the rate equation to decide which step could **not** be the rate-limiting step for this reaction and explain your reasoning. **[2 marks]**

4. Ozone reacts with nitrogen(IV) oxide to form oxygen and nitrogen(V) oxide.

$O_3(g) + 2NO_2(g) \longrightarrow O_2(g) + N_2O_5(g)$ The rate equation is rate = $k[O_3][NO_2]$

a) Explain why it is unlikely that this reaction proceeds via a single step. **[2 marks]**

b) Suggest a mechanism for this reaction. **[2 marks]**

c) Identify the rate-determining step in your mechanism. **[1 mark]**

Redox

Oxidation **I**s **L**oss, **R**eduction **I**s **G**ain of electrons.

Oxidation is an increase in oxidation number.

Reduction is a decrease in oxidation number.

An oxidising agent accepts electrons.

A reducing agent donates electrons.

Redox Reactions

$$Zn(s) + Cu^{2+}(aq) \longrightarrow Zn^{2+}(aq) + Cu(s)$$

The overall reaction can be separated into two half equations each showing only one species:

Zinc is oxidised. $Zn(s) \longrightarrow Zn^{2+}(aq) + 2e^-$
Copper is reduced. $Cu^{2+}(aq) + 2e^- \longrightarrow Cu(s)$

Zn is the reducing agent.
Cu^{2+} is the oxidising agent.

The reduced and oxidised forms of a substance are known as a redox pair, e.g. Cu^{2+}/Cu Zn^{2+}/Zn
ox/red ox/red

Electrochemical Cells

When an oxidising agent and a reducing agent are mixed together, electrons are transferred from the reducing agent to the oxidising agent and energy is released.

$$Zn(s) + Cu^{2+}(aq) \longrightarrow Zn^{2+}(aq) \qquad \Delta H = -217 \text{ kJ mol}^{-1}$$

If the two half reactions are separated physically but connected by a wire, as in an electrochemical cell, the released energy can be transferred to electrical energy.

In an electrochemical cell, one redox pair (Cu^{2+}/Cu) are in contact with each other within one half-cell and the other redox pair (Zn^{2+}/Zn) in the second half-cell.

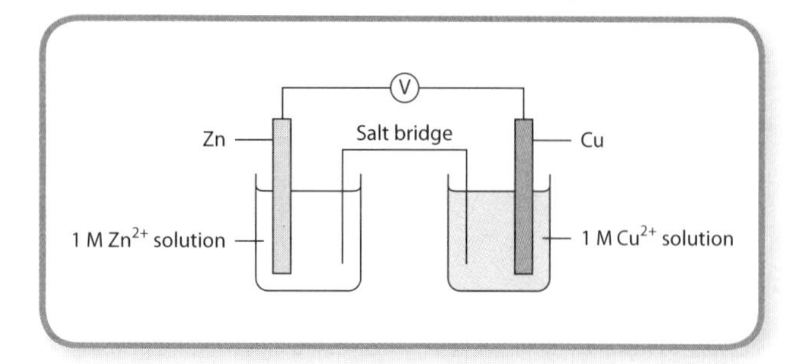

Electrons are transferred from the reducing agent to the oxidising agent via the wire.

The salt bridge completes the electrical circuit by allowing ions to migrate between the two half-cells.

If a high resistance voltmeter is added then electrons do not move but potential energy can be measured.

Standard Electrode Potentials

The tendency of a substance to be reduced is described by its standard electrode potential, $E^\ominus$.

The standard electrode potential compares the tendency of a substance to be reduced with the tendency of H^+ to be reduced in the standard hydrogen electrode under standard conditions.

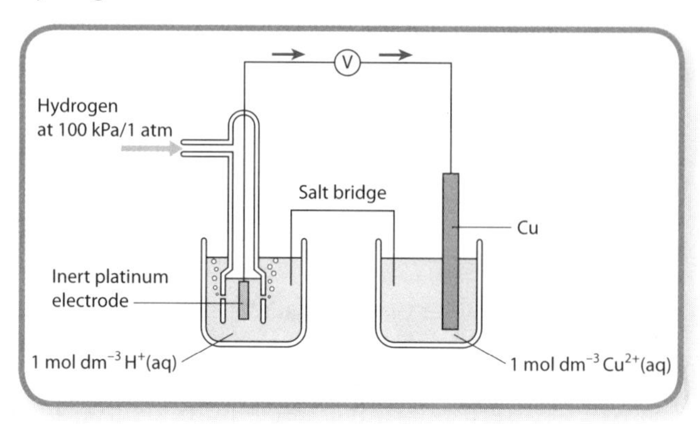

Standard conditions are:

- All solutions 1 mol dm^{-3}
- All gases 100 kPa/1 atm
- Temperature 298 K

The standard hydrogen electrode contains 1 mol dm^{-3} H^+ and has $H_2(g)$ bubbled through at 100 kPa/1 atm.

The reduction reaction for the hydrogen electrode is

$$2H^+(aq) + 2e^- \longrightarrow H_2(g)$$

This is defined as having a standard potential of 0 volts. The standard electrode potential of any half-cell connected to the hydrogen electrode is the reading on the voltmeter.

Half-cells with a positive electrode potential cause H_2 to be oxidised; those with a negative electrode potential cause H^+ to be reduced.

A half-cell can be a metal and aqueous ions, a gas and aqueous ions or two aqueous ions. When no metal is involved, an inert platinum electrode is submerged in the solution to transfer the electrons to or from the other half-cell. When one reactant is a gas, a gas electrode like the hydrogen electrode is used.

The Electrochemical Series

An electrochemical series lists the standard electrode potentials for redox pairs with the oxidised form as reactant and the reduced form as product.

Oxidising agent			Reducing agent	Standard potential $E^\ominus$ (V)
$K^+(aq)$	$+ e^-$	$\rightleftharpoons$	$K(s)$	−2.92
$Mg^{2+}(aq)$	$+ 2e^-$	$\rightleftharpoons$	$Mg(s)$	−2.38
$Zn^{2+}(aq)$	$+ 2e^-$	$\rightleftharpoons$	$Zn(s)$	−0.76
$Fe^{3+}(aq)$	$+ 3e^-$	$\rightleftharpoons$	$Fe(s)$	−0.04
$2H^+(aq)$	$+ 2e^-$	$\rightleftharpoons$	$H_2(g)$	0.00
$Cu^{2+}(aq)$	$+ 2e^-$	$\rightleftharpoons$	$Cu(s)$	+0.34
$Fe^{3+}(aq)$	$+ e^-$	$\rightleftharpoons$	$Fe^{2+}(aq)$	+0.77
$Ag^+(aq)$	$+ e^-$	$\rightleftharpoons$	$Ag(s)$	+0.80
$O_2(g) + 4H^+(aq)$	$+ 2e^-$	$\rightleftharpoons$	$2H_2O(l)$	+1.23
$MnO_4^-(aq) + 8H^+(aq)$	$+5e^-$	$\rightleftharpoons$	$Mn^{2+}(aq) + 4H_2O$	+1.49

A large positive $E^\ominus$ means that the reduced form of the redox pair is a good oxidising agent but the reduced form is a weak reducing agent.

$MnO_4^- + 8H^+ + 5e^- \rightleftharpoons Mn^{2+} + 4H_2O$ $E^\ominus$ +1.49 V

good oxidising agent poor reducing agent

A large negative $E^\ominus$ means that the reduced form of the redox pair is a poor oxidising agent but the reduced form is a good reducing agent.

$K^+ + e^- \rightleftharpoons K$ $E^\ominus$ −2.92 V

poor oxidising agent good reducing agent

Predicting the Direction of Redox Reactions from $E^\ominus$

When two half-cells are connected, the half-cell with the least positive $E^\ominus$ releases electrons to the electrode and an oxidation reaction occurs. The half-cell with the most positive $E^\ominus$ accepts electrons and a reduction occurs.

Cell Potential $E^\ominus_{cell}$ (emf)

The voltage when any two half-cells are connected is the cell potential $E^\ominus_{cell}$ (emf) and is the difference between the $E^\ominus$ of the two half-cells.

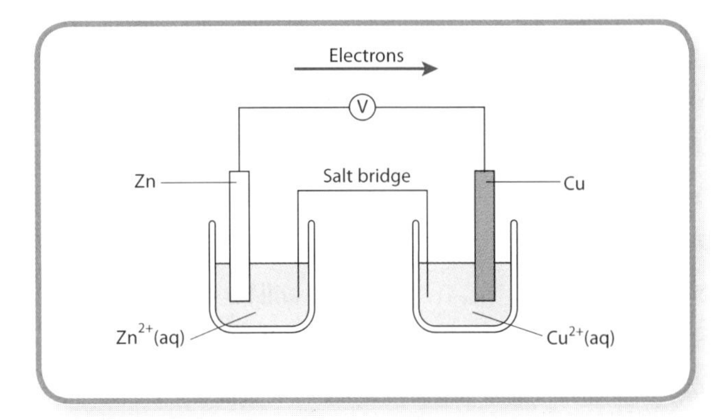

$Cu^{2+}(aq) + 2e^- \rightleftharpoons Cu(s)$ +0.34 V

$Zn^{2+}(aq) + 2e^- \rightleftharpoons Zn(s)$ −0.76 V

The half-cell with the most negative $E^\ominus$ will donate electrons to the half-cell with the most positive $E^\ominus$.

The Zn half-cell has the least positive $E^\ominus$.
Zn is oxidised. $Zn(s) \longrightarrow Zn^{2+}(aq) + 2e^-$
The electrons released travel through the wire to the copper half-cell. The Zn half-cell is the negative electrode.

The Cu half-cell has the most positive $E^\ominus$. It receives electrons from the electrode and Cu^{2+} is reduced.

$$Cu^{2+}(aq) + 2e^- \longrightarrow Cu(s)$$

The Cu half-cell is the positive electrode.

The difference in voltage between the two half-cells is the maximum voltage of the cell, $E^\ominus_{cell}$.

Most positive $E^\ominus$ − least positive $E^\ominus = E^\ominus_{cell}$

$E^\ominus_{cell} = +0.34 - (-0.76) = +1.10$ V

Examples of Calculating $E^\ominus_{cell}$ from $E^\ominus$

What is $E^\ominus_{cell}$ for the reaction between copper and silver half-cells?

$Cu^{2+}(aq)$	$+ 2e^-$	$\rightleftharpoons$	$Cu(s)$	+0.34 V
$Ag^+(aq)$	$+ e^-$	$\rightleftharpoons$	$Ag(s)$	+0.80 V

Most positive = Ag^+/Ag half-cell. This will receive electrons. Copper will supply electrons to silver ions.

$E^\ominus_{cell} = 0.8 - (+0.34) = +0.46\ V$

Note: Even though the number of electrons involved in the silver half-cell is only one while the copper half-cell supplies two electrons, the standard potential is not multiplied.

What is the $E^\ominus_{cell}$ for the reaction between zinc and magnesium half-cells?

$Mg^{2+}(aq)$	$+ 2e^-$	$\rightleftharpoons$	$Mg(s)$	−2.38 V
$Zn^{2+}(aq)$	$+ 2e^-$	$\rightleftharpoons$	$Zn(s)$	−0.76 V

Most positive = zinc. This half-cell will receive electrons. Magnesium gives electrons to zinc ions.

$E^\ominus_{cell} = -0.76 - (-2.38) = +1.62\ V$

Writing Cell Diagrams

The standard cell diagram can be represented using the following convention:

$$Pt(s)|H_2(g)|2H^+(aq)||Fe^{3+}(aq), Fe^{2+}(aq)|Pt(s)$$

The most negative half-cell is written on the left. The reduced form of the substance is written first then the oxidised form second.

If the reduced form and oxidised form are in different physical states, they are separated by a vertical line, e.g. $H_2(g)\,|\,2H^+(aq)$

If the oxidised and reduced forms are in the same state or phase, they are separated by a comma, e.g. $Fe^{2+}(aq)$, $Fe^{3+}(aq)$

If a platinum electrode is required this is written first followed by a vertical line, e.g. $Pt(s)\,|\,H_2(g)\,|\,2H^+(aq)$

A double vertical line represents the salt bridge.

$$Pt(s)|H_2(g)|2H^+(aq)||$$

The least negative electrode is written after the salt bridge. The oxidised form is written first and the reduced form second.

$$Pt(s)|H_2(g)|2H^+(aq)||Fe^{3+}(aq), Fe^{2+}(aq)$$

If a platinum electrode is required this is written last.

$$Pt(s)|H_2(g)|2H^+(aq)||Fe^{3+}(aq), Fe^{2+}(aq)\,|\,Pt(s)$$

Electrons are moving from the left-hand half-cell to the right-hand half-cell.

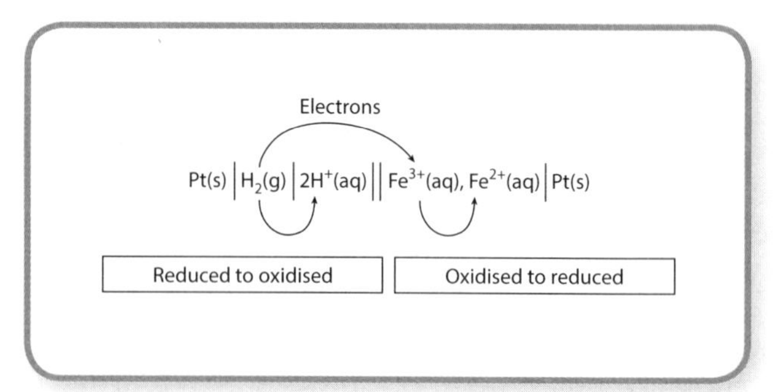

To the left of the salt bridge is the negative electrode; to the right of the salt bridge is the positive electrode.

$E_{cell} = E^\ominus$ right-hand electrode – $E^\ominus$ left-hand electrode

QUICK TEST

Use the table on page 45 to answer these questions.

1. Which is the best oxidising agent, Mg^{2+} or Ag^+?
2. Which is the best reducing agent, Mg or Ag?
3. If the Cu^{2+}/Cu half-cell was set against a standard hydrogen electrode, which would be the negative electrode?
4. Which species would be reduced if Fe^{2+}, Cu^{2+} and Ag^+ were mixed together?

PRACTICE QUESTIONS

Use the table on page 45 to answer these questions.

1. An electrochemical cell was set up using Ag^+/Ag and Fe^{3+}/Fe as the half-cells. E_{cell} was **[1 mark]**

   A +0.76 V ☐ **B** −0.76 V ☐

   C +0.84 V ☐ **D** −0.84 V ☐

2. Reducing agents that are able to reduce Cu^{2+} include **[1 mark]**

   A Zn, Ag and Fe^{2+} ☐ **B** Fe, H_2 and H_2O ☐

   C Mg, Fe^{2+} and Zn ☐ **D** Mg, Zn and H_2. ☐

3. The table shows the standard electrode potentials for some half-cell reactions.

Half equations			$E^\ominus$/V
$Cr^{3+}(aq) + 3e^-$	$\rightleftharpoons$	Cr(s)	−0.74
$Sn^{2+}(aq) + 2e^-$	$\rightleftharpoons$	Sn(s)	−0.14
$Cu^{2+}(aq) + 2e^-$	$\rightleftharpoons$	Cu(s)	+0.34
$I_2(aq) + 2e^-$	$\rightleftharpoons$	$2I^-(aq)$	+0.54
$Hg^{2+}(aq) + 2e^-$	$\rightleftharpoons$	Hg(l)	+0.85

   a) What is meant by standard electrode potential? **[3 marks]**

   b) Explain why the electrode potential for Cu^{2+}/Cu has a positive value while the electrode potential for Cr^{3+}/Cr has a negative value. **[1 mark]**

   c) Which is the strongest oxidising agent from the table above? **[1 mark]**

   d) Name two substances from the table that are able to reduce Cu^{2+}. **[2 marks]**

   The Cr^{3+}/Cr half-cell and Cu^{2+}/Cu half-cell were joined to make a cell and the E_{cell} value was measured.

   e) Write the conventional cell representation for this cell. **[2 marks]**

   f) Calculate the value of E_{cell} (emf) for this cell. **[1 mark]**

4. The standard electrode potential of a redox couple is found by setting it against a standard hydrogen electrode and finding the E_{cell} (emf).

   a) Draw a labelled diagram of the set up that would be used to measure $E^\ominus$ for $Fe^{2+} + 2e^- \rightleftharpoons Fe$. **[5 marks]**

   The value for $E^\ominus$ was found to be −0.41 V.

   b) In which direction do electrons flow in this cell? **[1 mark]**

   c) Write the reaction that occurs in the hydrogen half-cell when a current is flowing. **[1 mark]**

   d) What is the oxidising agent when a current is flowing? **[1 mark]**

Redox Equations

Predicting the Direction of Redox Reactions

Two redox couples/pairs have the potential to react together. The best oxidising agent will react with the best reducing agent.

Each redox pair has an electrode potential $E^\ominus$.

Example

Br_2/Br^- $E^\ominus$ +1.07 V

Fe^{3+}/Fe^{2+} $E^\ominus$ +0.77 V

The best oxidising agent is the oxidised form of the redox pair with the most positive $E^\ominus$. Br_2/Br^-

The best reducing agent is the reduced form of the redox pair with the least positive $E^\ominus$. Fe^{3+}/Fe^{2+}

Br_2 and Fe^{2+} react to form $2Br^-$ and Fe^{3+}.

If $E^\ominus_{cell}$ for two half-cells is positive then a reaction is feasible. $E^\ominus_{cell}$ predicts the thermodynamic feasibility of the reaction but not how fast the reaction will be. There may be no apparent reaction despite a positive $E^\ominus_{cell}$ because the reaction is too slow.

If $E^\ominus_{cell}$ for two half-cells is negative then the reaction is not feasible. This only applies to the standard conditions used to calculate $E^\ominus_{cell}$. Changing the concentrations and temperatures changes the electrode potential so reactions that appear to be unfeasible may be seen to happen.

Writing Ionic Equations from Half Equations

In a balanced redox equation the number of electrons provided by the reducing agent must match the number of electrons received by the oxidising agent.

To write the reaction equation:

1. Write out the half equation for the reduction reaction (most positive half-cell reaction).

   $Ag^+(aq) + e^- \longrightarrow Ag(s)$

2. Write out the half equation for the oxidation reaction. This is the $E^\ominus$ half equation in reverse.

   $Cu^{2+}(aq) + 2e^- \rightleftharpoons Cu(s)$

   Reversed to give
   $Cu(s) \longrightarrow Cu^{2+}(aq) + 2e^-$

3. If the number of electrons in the two half-cells is not the same, multiply the half equations until the number of electrons matches.

   Multiply Ag^+/Ag to give $2e^-$.
   $2Ag^+(aq) + 2e^- \longrightarrow 2Ag(s)$

4. Write the two half equations under one another and cancel out electrons and any other species that appear on both sides of the equation.

   $2Ag^+(aq) + \cancel{2e^-} \longrightarrow 2Ag(s)$

   $Cu(s) \longrightarrow + \cancel{2e^-}\ Cu^{2+}(aq)$

5. What remains is the ionic equation for the reaction:

   $$Cu(s) + 2Ag^+(aq) \longrightarrow Cu^{2+}(aq) + 2Ag(s)$$

Half equations that contain compound ions, hydrogen ions and water look more complicated but the equation is written in exactly the same way, e.g. the reaction between manganate ions and iron(II) ions in acidic solution.

$Fe^{3+} + e^- \rightleftharpoons Fe^{2+}$ +0.77 V

$MnO_4^- + 8H^+ + 5e^- \rightleftharpoons Mn^{2+} + 4H_2O$ +1.49 V

The Fe^{3+}/Fe^{2+} half-cell has the least positive electrode potential so Fe^{2+} will be oxidised to Fe^{3+}. MnO_4 will be reduced to Mn^{2+}.

1. Oxidation reaction:
   $MnO_4^- + 8H^+ + 5e^- \longrightarrow Mn^{2+} + 4H_2O$

2. Reduction reaction:
   $Fe^{2+} \longrightarrow Fe^{3+} + e^-$

3. Multiply equations so electrons match:
   $5Fe^{2+} \longrightarrow 5Fe^{3+} + 5e^-$

4. Write out both equations and cancel down:
   $MnO_4^- + 8H^+ + \cancel{5e^-} \longrightarrow Mn^{2+} + 4H_2O$
   $5Fe^{2+} \longrightarrow 5Fe^{3+} + \cancel{5e^-}$

5. Overall reaction:
   $MnO_4^- + 8H^+ + 5Fe^{2+} \longrightarrow Mn^{2+} + 4H_2O + 5Fe^{3+}$

Writing and Balancing Half Equations

If the redox pair is known then the half equation can be written. The atoms and charges on either side of a half equation must balance. Also, only one species should change oxidation state.

For monatomic ions the charges are balanced with electrons, e.g. writing the half equation for

Fe^{3+}/Fe^{2+} $\quad Fe^{3+} \longrightarrow Fe^{2+}$

There is one more positive charge on the left-hand side. Balance by adding one electron to the left-hand side.

$$Fe^{3+} + e^- \longrightarrow Fe^{2+}$$

For compound ions the atoms are balanced using H_2O and H^+, e.g.

$$Cr_2O_7^{2-} \longrightarrow Cr^{3+}$$

1. Balance the atoms of the oxidised and reduced element.

$$Cr_2O_7^{2-} \longrightarrow 2Cr^{3+}$$

2. Add water molecules on the right to balance the oxygen atoms.

$$Cr_2O_7^{2-} \longrightarrow 2Cr^{3+} + 7H_2O$$

3. Add hydrogen ions on the left to balance the hydrogen atoms.

$$Cr_2O_7^{2-} + 14H^+ \longrightarrow 2Cr^{3+} + 7H_2O$$

4. Add electrons to balance the charges.

$$Cr_2O_7^{2-} + 14H^+ + 6e^- \longrightarrow 2Cr^{3+} + 7H_2O$$

This is only true when the reaction takes place in acidic solution.

For reactions in alkaline solution:

1. Balance the equation as if it were in acidic conditions by adding H_2O and H^+.
2. Add enough OH^- to the H^+ side of the equation to neutralise all the H^+.
3. Add the same number of OH^- to the other side of the equation.
4. Cancel down all species that appear on both sides of the equation.

Electrode Potentials and Disproportionation Reactions

In a disproportionation reaction the same species is both oxidised and reduced, e.g.

$$Cu_2O + H_2SO_4 \longrightarrow Cu + CuSO_4 + H_2O$$

Oxd'n state of Cu $+1$ $\quad$ 0 $\quad$ $+2$

This is possible when:

- the same species appears as an oxidising agent and a reducing agent in different half equations
- $E^\ominus$ for the half equation where it is the oxidising agent is more positive than $E^\ominus$ for the half equation where it is the reducing agent.

Cu^+ will disproportionate since

$Cu^{2+} + e^- \longrightarrow Cu^+$ $\quad E^\ominus = +0.15\text{ V}$ $\quad Cu^+$ as reducing agent

$\mathbf{Cu^+} + e^- \longrightarrow Cu$ $\quad E^\ominus = +0.52\text{ V}$ $\quad$ **Cu^+** as oxidising agent

$E^\ominus$ of Cu^+ as oxidising agent is more positive than $E^\ominus$ Cu^+ as reducing agent.

When Cu^+ is oxidised to Cu^{2+} it releases electrons to reduce Cu^+ to Cu.

Mn^{2+} will not disproportionate:

$Mn^{2+} + 2e^- \longrightarrow Mn$ $\quad E^\ominus = -1.19\text{ V}$ $\quad Mn^{2+}$ as oxidising agent

$MnO_2 + 4H^+ + 2e^- \longrightarrow Mn^{2+} + H_2O$ $\quad E^\ominus = +1.23\text{ V}$ $\quad Mn^{2+}$ as reducing agent

$E^\ominus$ of Mn^{2+} as oxidising agent is less positive than $E^\ominus$ of Mn^{2+} as reducing agent.

MnO_4^{2-} will disproportionate in acid solution:

$MnO_4^- + e^- \longrightarrow MnO_4^{2-}$ $\quad E^\ominus = +0.56\text{ V}$

$MnO_4^{2-} + 4H^+ + 2e^- \longrightarrow MnO_2 + 2H_2O$ $\quad E^\ominus = +1.70\text{ V}$

$$3MnO_4^{2-} + 4H^+ \longrightarrow 2MnO_4^- + MnO_2 + 2H_2O$$

Repeated Redox Reactions with the Same Reagent

Sometimes the product of one reduction reaction can be reduced again by the same reducing agent.

Vanadium has many stable oxidation states and each has its own colour.

Oxidation state	+2	+3	+4	+5
	V^{2+}	V^{3+}	VO^{2+}	VO_2^+
Colour	purple	green	blue	yellow

$Zn^{2+} + 2e^- \rightleftharpoons Zn \quad E^\ominus = -0.76\ V$

$V^{3+} + e^- \rightleftharpoons V^{2+} \quad E^\ominus = -0.26\ V$

$VO^{2+} + 2H^+ + e^- \rightleftharpoons V^{3+} + H_2O \quad E^\ominus = +0.34\ V$

$VO_2^+ + 2H^+ + e^- \rightleftharpoons VO^{2+} + H_2O \quad E^\ominus = +1.00\ V$

Zn is a stronger reducing agent than all vanadium ions. If excess Zn is added to a solution of VO_2^+ it reduces each vanadium ion in turn.

	Colour
1. $VO_2^+ + 2H^+ + Zn \longrightarrow \mathbf{VO^{2+}} + H_2O + Zn^{2+}$	blue
2. $VO^{2+} + H_2O + Zn \longrightarrow \mathbf{V^{3+}} + H_2O + Zn^{2+}$	green
3. $V^{3+} + Zn \longrightarrow \mathbf{V^{2+}} + Zn^{2+}$	purple

Note: The colour changes seen are yellow ⟶ green (a mixture of the yellow and blue ions) ⟶ blue ⟶ green (a different ion) ⟶ purple

The Sequence of Redox Reactions

The larger the value of $E^\ominus_{cell}$, the greater the increase in entropy. In a mixture of oxidising and reducing agents the first reactions to occur will be between those that give the biggest $E^\ominus_{cell}$.

Example

Which reaction will occur if zinc, silver ions and copper ions are together in solution?

$Zn^{2+}(aq) + 2e^- \rightleftharpoons Zn(s) \quad E^\ominus = -0.76\ V$

$Cu^{2+}(aq) + 2e^- \rightleftharpoons Cu(s) \quad E^\ominus = +0.34\ V$

$Ag^+(aq) + e^- \rightleftharpoons Ag(s) \quad E^\ominus = +0.80\ V$

The Ag^+ will be reduced to Ag first. The Cu^{2+} ions will only be reduced if there is excess Zn.

In practice this is not always what is observed since it is only true under standard conditions.

$E^\ominus_{cell}$ and Entropy

The free energy, ΔG, of a redox reaction is proportional to the negative of $E^\ominus_{cell}$ for that reaction.

$$\Delta G \propto -E^\ominus_{cell}$$

A positive $E^\ominus_{cell}$ means a negative ΔG, and so a feasible reaction. As the value of $E^\ominus_{cell}$ becomes more positive, ΔG becomes more negative.

The total entropy change, ΔS_{tot}, for a redox reaction is directly proportional to $E^\ominus_{cell}$.

$$\Delta G \propto \Delta S_{tot}$$

A positive $E^\ominus_{cell}$ means a positive ΔS_{total}, and so a feasible reaction. As the value for $E^\ominus_{cell}$ becomes more positive, so does the total entropy change.

$E^\ominus_{cell}$ and Equilibria

Under standard conditions, $E^\ominus_{cell}$ is directly proportional to the logarithm of the equilibrium constant.

$$E^\ominus_{cell} \propto \ln K$$

A large equilibrium constant means a large $E^\ominus_{cell}$ and a feasible reaction.

An equilibrium constant of less than 1 means that lnK is negative, $E^\ominus_{cell}$ is negative and the reaction is not feasible in this direction.

QUICK TEST

1. State a reason why a redox reaction with a positive E_{cell} may not happen in the laboratory.
2. Write the equation for the reaction between Zn and V^{3+}.
3. Write the half equation for the reduction of NO_3^- to NO and water.
4. Does VO^{2+} disproportionate?
   $VO^{2+} + 2H^+ + e^- \rightleftharpoons V^{3+} + H_2O \quad E^\ominus = +0.34\ V$
   $VO_2^+ + 2H^+ + e^- \rightleftharpoons VO^{2+} + H_2O \quad E^\ominus = +1.00\ V$
5. What will be the final colour of solution when V^{2+} and VO^{2+} are mixed in acidic conditions?
6. If $E^\ominus_{cell}$ is positive, what does this tell us about ΔG under standard conditions?

PRACTICE QUESTIONS

1. $MnO_4^-(aq)$ is added to a solution of $FeSO_4$ and a redox reaction occurs.

$MnO_4^- + 8H^+ + 5e^- \rightleftharpoons Mn^{2+} + 4H_2O \quad E^\ominus = +1.51\ V$

$Fe^{3+} + e^- \rightleftharpoons Fe^{2+} \quad E^\ominus = +0.77\ V$ **[1 mark]**

A The oxidising agent is Fe^{3+}. ☐ **B** The reducing agent is Fe^{2+}. ☐

C The oxidising agent is Mn^{2+}. ☐ **D** The reducing agent is MnO_4^-. ☐

2. HClO can be reduced to Cl_2 in acidic solution. The half equation for this reaction is **[1 mark]**

A $HClO + H^+ + e^- \rightleftharpoons Cl^- + H_2O$ ☐ **B** $HClO + H^+ + e^- \rightleftharpoons Cl_2 + H_2O$ ☐

C $HClO + H^+ + e^- \rightleftharpoons 0.5Cl_2 + H_2O$ ☐ **D** $2HClO + 2e^- \rightleftharpoons Cl_2 + H_2O$ ☐

3. Cans used in preserving foods can be made from steel plated with a thin layer of tin. The tin acts as a barrier to prevent the iron in the steel from corroding as a result of reacting with chemicals in the food.

	$E^\ominus$/V
$Fe^{2+}(aq) + 2e^- \longrightarrow Fe(s)$	−0.41
$Sn^{2+}(aq) + 2e^- \longrightarrow Sn(s)$	−0.14
$2H^+(aq) + 2e^- \longrightarrow H_2(g)$	0.00
$NO_3^-(aq) + 2H^+ + 2e^- \longrightarrow NO_2^-(aq) + H_2O(l)$	+0.94

Food scientists have found that some tomatoes contain high levels of nitrates as a result of fertiliser use. When these tomatoes are canned, the tin coating of the can quickly disintegrates, exposing the iron to the tomatoes. The tomatoes are naturally acidic.

a) Why does the presence of nitrates in the can cause the tin to disintegrate? **[3 marks]**

b) Write a balanced equation for the reaction between tin and nitrates. **[1 mark]**

c) What value for $E^\ominus_{cell}$ would be obtained if the Sn^{2+}/Sn half-cell were set against the standard hydrogen electrode? **[1 mark]**

d) Suggest a reason why there is no reaction between Sn and the acid in the tomatoes. **[1 mark]**

e) Once the layer of tin has been removed the can rapidly corrodes. Explain why. **[1 mark]**

Scientists are also very concerned about the presence of nitrites (nitrate(III), NO_2^-) in canned foods. Nitrites can be reduced to ammonia in the presence of hydrogen ions.

f) Write a balanced half equation for the reduction of NO_2^- to NH_3 in acidic solution. **[2 marks]**

4. Solid iodine was added to an acidified solution of V^{2+}.

			$E^\ominus$/V
$I_2 + 2e^-$	$\rightleftharpoons$	$2I^-$	+0.54
$V^{3+} + e^-$	$\rightleftharpoons$	V^{2+}	−0.26
$VO^{2+} + 2H^+ + e^-$	$\rightleftharpoons$	$V^{3+} + H_2O$	+0.34
$VO_2^+ + 2H^+ + e^-$	$\rightleftharpoons$	$VO^{2+} + H_2O$	+1.00

a) Give the final oxidation state of the vanadium ion that results. **[1 mark]**

b) Write a balanced equation for the reaction. **[2 marks]**

Storing Electricity/Rusting

Storing Electricity

Dry Cells

In a dry cell battery the two half-cells are inserted one inside the other, e.g alkaline batteries.

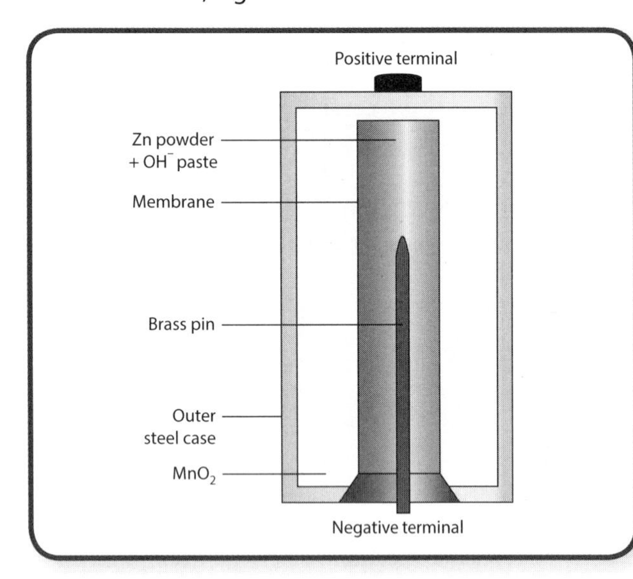

An aqueous paste is used in place of a solution. The outer steel case acts as one electrode. The inner brass rod is the other electrode. The outer half-cell contains MnO_2; the inner half-cell contains zinc powder in a OH^- gel. The two half-cells are separated by a membrane that prevents them from mixing but allows ions to pass across.

The zinc from the inner cell is oxidised and releases electrons to the external circuit.

$$Zn \longrightarrow Zn^{2+} + 2e^-$$

The MnO_2 in the outer half-cell is reduced by electrons from the positive terminal.

$$2MnO_2 + H_2O + 2e^- \longrightarrow Mn_2O_3 + 2OH^-$$

Overall reaction:

$$2MnO_2 + H_2O + Zn \longrightarrow Mn_2O_3 + 2OH^- + Zn^{2+}$$

The cell is 'flat' when there is no longer sufficient zinc. It must then be disposed of.

Rechargeable Batteries

Rechargeable batteries work in the same way as non-rechargeable batteries. It is possible to reverse the redox reaction by using an external direct current power source to move electrons in the opposite direction.

When power is being drawn from the cell, the negative electrode releases electrons to the circuit and an oxidation reaction occurs. A reduction reaction occurs at the positive terminal.

When the cell is being charged the reverse happens. Oxidation occurs at the positive electrode and reduction at the negative electrode, e.g. nickel–cadmium batteries.

When discharging (providing a current):

$$NiO_2(s) + 2H_2O(l) + 2e^- \longrightarrow Ni(OH)_2(s) + 2OH^-(aq)$$
$$Cd(s) + 2OH^-(aq) \longrightarrow Cd(OH)_2 + 2e^-$$

Overall reaction:

$$NiO_2(s) + 2H_2O(l) + Cd(s) \longrightarrow Ni(OH)_2(s) + Cd(OH)_2(s)$$

When recharging, the reverse reaction occurs:

$$Ni(OH)_2(s) + Cd(OH)_2(s) \longrightarrow NiO_2(s) + 2H_2O(l) + Cd(s)$$

Advantages: can be used many times before having to be replaced.

Disadvantages: Cadmium is a toxic heavy metal so production and disposal of nickel–cadmium batteries poses a potential environmental hazard.

Lithium Cells

Lithium cobalt cells are commonly used in mobile phones and laptops. They are compact with a high energy density.

Reaction at the positive electrode during discharge:

$$Li^+ + CoO_2 + e^- \longrightarrow Li[CoO_2]$$

The reverse occurs during charging.

Reaction at the negative electrode during discharge:

$$Li \longrightarrow Li^+ + e^-$$

The reverse occurs during charging.

Overall reaction: $CoO_2 + Li \longrightarrow Li[CoO_2]$

Disadvantages: Lithium cobalt cells contain an organic electrolyte, which is flammable. The combination of chemicals in the cell means that it is possible for the cell to overheat and catch fire.

Fuel Cells

In a fuel cell the reactants are continuously replaced.

The reaction is between a fuel and oxygen and the energy of the reaction is transferred as electrical energy rather than in combustion.

The Hydrogen Fuel Cell

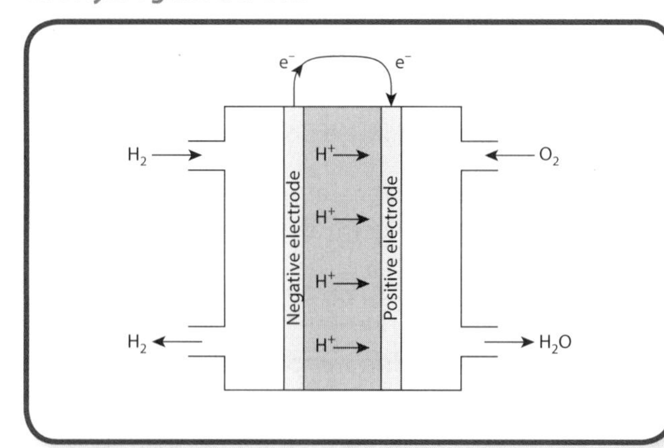

- Hydrogen molecules enter the cell and come into contact with the negative electrode. (The electrodes contain a transition metal catalyst such as Pt.)
- They are oxidised to H^+:

$$H_2 \longrightarrow 2H^+ + 2e^-$$

- Electrons leave the cell via the negative electrode and hydrogen ions pass across the porous electrode and membrane into the central section of the cell, which contains an acid electrolyte.
- H^+ migrates across to the positive electrode.
- Electrons enter the cell via the positive electrode.
- Oxygen molecules enter the cell and come into contact with the positive electrode. They react with H^+ and electrons and are reduced to water.

$$O_2 + 4H^+ + 4e^- \longrightarrow 2H_2O$$

Overall cell reaction: $2H_2 + O_2 \longrightarrow 2H_2O$

Alkaline Hydrogen Fuel Cell

Alkaline fuel cells use an alkaline electrolyte and it is OH^- that migrates across the central section.

- Oxygen molecules enter the cell and come into contact with the positive electrode.
- They pick up electrons and react with water, and are reduced to hydroxide ions:

$$O_2 + 2H_2O + 4e^- \longrightarrow 4OH^-$$

- Hydroxide ions migrate across the central section to the negative electrode.
- Hydrogen molecules enter the cell and come into contact with the negative electrode. They react with OH^-, and oxidise and release electrons to the electrode.

$$H_2 + 2OH^- \longrightarrow 2H_2O + 2e^-$$

Overall cell reaction: $2H_2 + O_2 \longrightarrow 2H_2O$

Benefits and Drawbacks of Hydrogen Cells

Benefits:

- Hydrogen is a renewable resource.
- The products are non-toxic and do not contribute to global warming (water is a greenhouse gas but release of water vapour by fuel cells will not increase atmospheric water vapour).
- They are very effective in generating energy from hydrogen.

Drawbacks:

- An alternative source of energy is needed to generate the hydrogen, usually by electrolysis of water.
- Hydrogen is flammable and must be stored under pressure, so presents a hazard.
- The cost of transferring from conventional energy sources to fuel cells is high.

Fuel Cells Other Than Hydrogen Cells

Hydrogen-rich organic molecules such as alcohols can be used as the fuel in cells, e.g. ethanol.

The oxidation half-cell for ethanol as a fuel:

$$CH_3CH_2OH + H_2O \longrightarrow CH_3COOH + 4H^+ + 4e^-$$

The reduction half-cell: $O_2 + 4H^+ + 4e^- \longrightarrow 2H_2O$

Overall cell reaction $C_2H_5OH + O_2 \longrightarrow CH_3CO_2H + H_2O$

Ethanol can be produced by fermentation, a relatively low carbon technology.

Methanol can be oxidised completely by the fuel cell and generates large amounts of energy.

$$CH_3OH + 1.5O_2 \longrightarrow CO_2 + 2H_2O$$

Advantages of methanol include: it can be generated from general biomass and so does not use up food sources in the way that ethanol from fermentation does; it is very energy dense compared with other fuels.

Problems with methanol fuel cells include: the toxicity and cost of production of methanol; it produces a greenhouse gas.

Rusting

Rusting of iron is an electrochemical process.

The two half reactions involved are:

$$Fe \longrightarrow Fe^{2+} + 2e^-$$
$$O_2 + 2H_2O + 4e^- \longrightarrow 4OH^-$$

Example
When water collects in a scratch on the surface of an iron object.

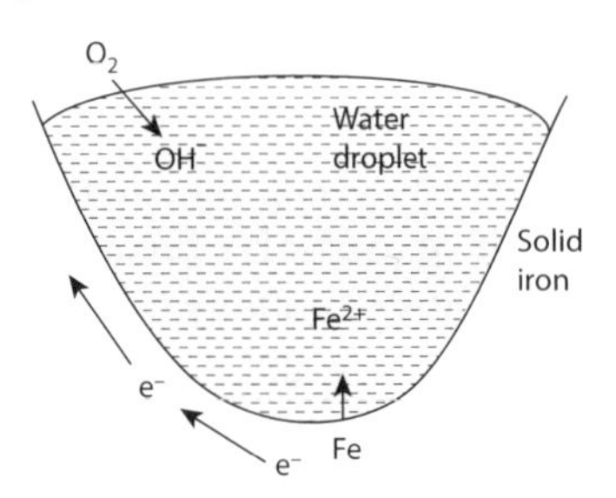

Oxygen from the air at the edge of the water droplet reacts with water and electrons from the iron.

$$O_2 + 2H_2O + 4e^- \longrightarrow 4OH^-$$

The hydroxide ions diffuse into the water.

This is the reduction half reaction and it occurs at the place of highest oxygen concentration.

Electrons move through the iron to replace those removed by the oxygen reaction, leaving Fe^{2+} at the point in the metal that is furthest from the oxygen supply, the centre of the drop of water.

$Fe \rightarrow Fe^{2+} + 2e^-$ This is the oxidation reaction. It occurs at the place of lowest oxygen concentration.

The $Fe^{2+}(aq)$ diffuses into the water.

The OH^- and Fe^{2+} meet in the water and form insoluble $Fe(OH)_2(s)$. This is further oxidised by oxygen in the air to form $Fe(OH)_3.XH_2O(s)$, which is rust.

Prevention of Rusting

One way of preventing rusting is to place an impermeable barrier between the iron and the water, such as painting or coating in polymer.

An alternative is to put a metal with a more negative electrode potential in electrical contact with the iron. This is known as sacrificial protection, as this more reactive metal is sacrificed for the iron.

For example blocks of zinc are attached to the steel hull of boats. When iron from the steel is oxidised, electrons are donated from the zinc metal $Zn \longrightarrow Zn^{2+} + 2e^-$ and travel through the metal hull to reduce the Fe^{2+} back to Fe.

QUICK TEST

1. Why do alkaline batteries only last for a limited time?
2. If the half-cell reaction when a rechargeable battery is discharging is $Cd + 2OH^- \longrightarrow Cd(OH)_2 + 2e^-$, what is the half reaction when it is being charged?
3. Why are nickel–cadmium batteries a potential environmental hazard?
4. Give an advantage of lithium cells over zinc alkaline batteries.
5. What is the overall reaction when a hydrogen fuel cell is producing current?
6. Why is water needed for the rusting process?

PRACTICE QUESTIONS

1. In one kind of alkaline battery the overall cell reaction is $2MnO_2 + H_2O + Zn \longrightarrow Mn_2O_3 + 2OH^- + Zn^{2+}$.
The reaction at the positive terminal is **[1 mark]**

A $Zn \longrightarrow Zn^{2+} + 2e^-$ ☐ **B** $2MnO_2 \longrightarrow Mn_2O_3$ ☐

C $Zn + H_2O \longrightarrow Zn(OH)_2 + 2e$ ☐ **D** $2MnO_2 + H_2O + 2e^- \longrightarrow Mn_2O_3 + 2OH^-$ ☐

2. A hydrogen fuel cell can be used for transport vehicles. Which is true? **[1 mark]**

A It must be continually supplied with H^+. ☐

B The only waste product during use is water. ☐

C It uses air as a fuel supply. ☐

D It requires regular recharging. ☐

3. The basic principle of the hydrogen fuel cell can be applied to the oxidation of other fuels such as methanol. The overall reaction in a methanol fuel cell is the same as if the methanol were burned but it is a more effective way of 'capturing' the energy. The reaction at the positive electrode is the same as in the acid hydrogen fuel cell: $O_2 + 4H^+ + 4e^- \longrightarrow 2H_2O$

a) Write the overall reaction equation for a methanol fuel cell. **[1 mark]**

b) Write the half equation for the reaction at the negative electrode. Methanol reacts with water in this reaction. **[2 marks]**

Despite the efficiency of the methanol fuel cell, more attention has been paid to the far less efficient ethanol fuel cell.

c) Suggest two advantages of ethanol fuel cells over methanol fuel cells. **[2 marks]**

d) Give reasons why some people consider that the methanol fuel cell is no more carbon neutral than a conventional internal combustion engine. **[2 marks]**

4. The metal hull of boats is often protected by attaching blocks of zinc at points on the hull.

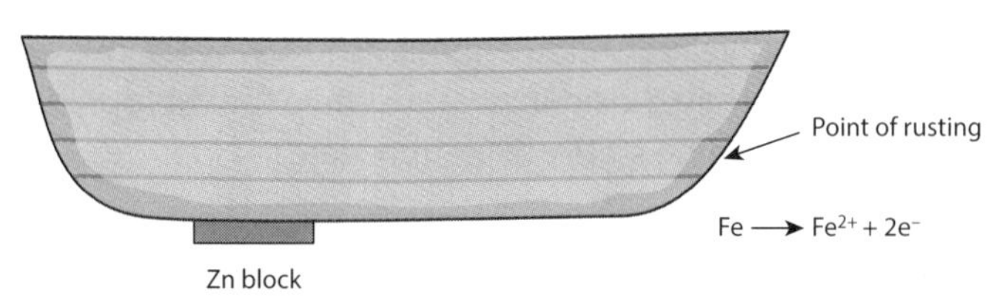

a) What reaction is occurring at the zinc block? **[1 mark]**

b) Explain how the zinc is able to protect the boat from rusting at the point shown. Copy the diagram and show the movement of electrons and ions. State any relevant equations. **[3 marks]**

c) Suggest what might happen if a copper plate were attached to the hull of the boat. Give a reason for your suggestion. **[2 marks]**

$Cu^{2+} + 2e^- \rightleftharpoons Cu$ $E^\ominus$ +0.34 V $Fe^{2+} + 2e^- \rightleftharpoons Fe$ $E^\ominus$ −0.44 V $Zn^{2+} + 2e^- \rightleftharpoons Zn$ $E^\ominus$ −0.76 V

d-block Elements

Electronic Configuration

d-block elements have their highest energy electron in a d-orbital.

Since the energy of the 4s sub-shell is lower than that of the 3d sub-shell, all the d-block elements in Period 4 have electrons in the 4s sub-shell.

Sc [Ar] $3d^1$ $4s^2$
Ti [Ar] $3d^2$ $4s^2$
V [Ar] $3d^3$ $4s^2$
Cr [Ar] $3d^5$ $4s^1$
Mn [Ar] $3d^5$ $4s^2$
Fe [Ar] $3d^6$ $4s^2$
Co [Ar] $3d^7$ $4s^2$
Ni [Ar] $3d^8$ $4s^2$
Cu [Ar] $3d^{10}$ $4s^1$
Zn [Ar] $3d^{10}$ $4s^2$

Cr and Cu have anomalous electron structures.

Predicted filling order for Cr [Ar] $4s^2$ $3d^4$
Observed filling order Cr [Ar] $3d^5$ $4s^1$

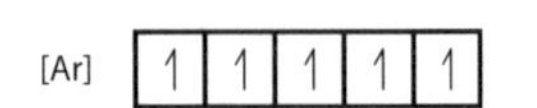

The lowest energy position is when each 4s and 3d orbital contains one electron.

Predicted filling order for Cu [Ar] $4s^2$ $3d^9$
Observed filling order Cu [Ar] $4s^1$ $3d^{10}$

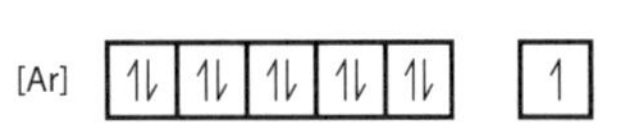

The lowest energy position is when 10 electrons fill the 3d sub-shell and the third electron shell is completely full.

Transition Elements

Transition elements can form at least one stable ion that has an incomplete d sub-shell.

All of the Period 4 d-block elements are transition elements except Sc and Zn.

Note: If you are studying with AQA the definition includes atoms as well as ions, which means Sc **is** a transition element.

Sc forms only one ion, Sc^{3+}, with an empty d sub-shell.

Sc^{3+} $1s^2$ $2s^2$ $2p^6$ $3s^2$ $3s^2$ $3p^6$

Zn forms only one ion, Zn^{2+}, with a completely full d sub-shell.

Zn^{2+} $1s^2$ $2s^2$ $2p^6$ $3s^2$ $3s^2$ $3p^6$ $3d^{10}$

When d-block elements form ions, the first electrons to be lost are the outer s sub-shell electrons.

Transition Metals Have Variable Oxidation States

Stable oxidation states of Period 4 d-block elements

	+1	+2	+3	+4	+5	+6	+7
Sc			Sc^{3+}				
Ti			Ti^{3+}	TiO^{2+}			
V		V^{2+}	V^{3+}	VO^{2+}	VO_2^+		
Cr		Cr^{2+}	Cr^{3+}			$Cr_2O_7^-$	
Mn		Mn^{2+}		MnO_2			MnO_4^-
Fe		Fe^{2+}	Fe^{3+}				
Co		Co^{2+}	Co^{3+}				
Ni		Ni^{2+}					
Cu	Cu^+	Cu^{2+}					
Zn		Zn^{2+}					

Successive ionisation energies show why Ca ions only exist in the +2 oxidation state while the d-block ions have variable stable oxidation states.

Ionisation enthalpy kJ mol^{-1}				
	$\Delta H^\ominus_{IE}(1)$	$\Delta H^\ominus_{IE}(2)$	$\Delta H^\ominus_{IE}(3)$	$\Delta H^\ominus_{IE}(4)$
Ca	+596	+1152	+4918	+6480
V	+656	+1420	+2834	+4513

Ca, an s-block element, shows a large increase in energy required for the third ionisation enthalpy. The third electron is in a completed inner 3p sub-shell. This makes oxidation states higher than +2 very unlikely.

The energy of 4s and 3d sub-shells is very similar. V shows a steady increase in ionisation enthalpy for each subsequent electron. For d-block elements there is no big energy jump in removing a third or subsequent electron from the d sub-shell. This means that several stable oxidation states are possible.

Transitions Elements as Catalysts

Heterogeneous catalysts are in a different phase to the reactants, often a solid catalyst and gaseous reactants. The reaction occurs on the surface of the catalyst.

The principal stages of the reaction are:

- Reactants adsorb (bond to the surface) to the catalyst, which weakens the bonds in the reactants and holds the reactants in a favourable orientation.
- Bonds in the reactants break.
- New bonds form in the products.
- The product desorbs and diffuses away from the catalyst.

Transition metals and their compounds often make good heterogeneous catalysts because of the availability of the d and s orbital electrons, which can form weak bonds with the reactants. Bonds must be strong enough to hold the reactants but weak enough to allow the products to desorb.

Examples of heterogeneous catalysts:

- Pt/Rh in a car catalytic converter.
- Ni in the hydrogenation of vegetable oils.
- Fe in the manufacture of NH_3 by the Haber process.
- MnO_2 in the decomposition of H_2O_2.

Homogeneous catalysts are in the same phase as the reactants, often both in solution. Transition metals make good catalysts for redox reactions because of their variable oxidation states.

The principal stages of the reaction are:

- One reactant is oxidised while the catalyst is reduced. This gives one product of the reaction and a reduced catalyst.
- The second reactant is reduced while the catalyst is oxidised. This gives the second product and the catalyst returns to its original oxidation state.

(If the catalyst is in its lowest oxidation state then the two stages occur in the reverse order.)

Example

Fe^{2+} catalyses the iodide/peroxodisulfate reaction

$$2I^-(aq) + S_2O_8^{2-}(aq) \longrightarrow 2SO_4^{2-}(aq) + I_2(aq)$$

Stage 1 $2Fe^{2+} + S_2O_8^{2-} \longrightarrow 2Fe^{3+} + 2SO_4^{2-}$

Stage 2 $2Fe^{3+} + 2I^- \longrightarrow 2Fe^{2+} + I_2$

The uncatalysed reaction has a high activation energy because the reactants both have negative charges and therefore repel each other making successful collisions difficult.

V_2O_5 is a Catalyst in the Contact Process

Vanadium(V) oxide is a heterogeneous catalyst that changes oxidation state when it catalyses the oxidation of sulfur dioxide to sulfur trioxide in the first step of sulfuric acid production.

$$2SO_2(g) + O_2(g) \rightleftharpoons 2SO_3(g)$$

Stage 1 $2V_2O_5 + 2SO_2 \longrightarrow 2SO_3 + 2V_2O_4$

Stage 2 $2V_2O_4 + O_2 \longrightarrow 2V_2O_5$

Autocatalysis

Mn^{2+} is both product and catalyst for the reaction between manganate(VII) and ethandioate. MnO_4^- is often used in redox titrations. When it is reduced from MnO_4^- to Mn^{2+} there is a colour change from pink to colourless.

$$2MnO_4^- + 16H^+ + 5C_2O_4^{2-} \longrightarrow 2Mn^{2+} + 10CO_2 + 8H_2O$$

At the start of the titration the pink colour takes a long time to fade (the $C_2O_4^{2-}$ must be warmed to speed up the titration). Once some Mn^{2+} has been formed, the pink colour of added MnO_4^- begins to disappear instantly until the end point is reached.

Redox Titrations

A redox titration can be used to measure the quantity of an oxidising or reducing agent.

To measure a reducing agent, titrate with a known concentration of oxidising agent.

Oxidising solution	Half equation	End point detection
$KMnO_4$ (intense purple)	$MnO_4^- + 8H^+ + 5e^- \longrightarrow Mn^{2+} + 4H_2O$	when the first purple colour is seen
I_2 (brown)	$I_2 + 2e^- \longrightarrow 2I^-$	add starch indicator, which turns blue-black at end point
$K_2Cr_2O_7$ (orange)	$Cr_2O_7^{2-} + 14H^+ + 6e^- \longrightarrow 2Cr^{3+} + 7H_2O$	add redox indicator, which changes colour at end point

If the reaction equation is not given then use the half equations to construct the equation. Calculate the moles of oxidising agent used in the titration, then multiply by the mole ratio in the equation to find the moles of the reducing agent.

A common way of measuring the concentration of an oxidising agent is to add an excess of iodide ions. This reduces the substance being measured and oxidises the iodide to iodine: $2I^- \longrightarrow I_2 + 2e^-$

The amount of iodine produced is proportional to the amount of oxidising agent being measured.

The quantity of iodine produced is then determined by titrating against a known concentration of thiosulfate.

$$I_2 + 2S_2O_3^{2-} \longrightarrow 2I^- + S_4O_6^{2-}$$

Starch is added close to the endpoint, which gives a blue-black colour. The end point is when the colour disappears.

Example

2.10 g of impure copper was dissolved in nitric acid and made up to 250 cm^3.

Equation 1: $Cu \longrightarrow Cu^{2+} + 2e^-$

25.0 cm^3 was removed and added to 25.0 cm^3 of 0.3 mol dm^{-3} KI(aq), which was an excess of KI.

Equation 2: $2Cu^{2+} + 4I^- \longrightarrow 2CuI + I_2$

The iodine was titrated against 0.100 mol dm^{-3} $Na_2S_2O_3$(aq).

Equation 3: $I_2 + 2S_2O_3^{2-} \longrightarrow 2I^- + S_4O_6^{2-}$

The average titre was 24.50 cm^3. What was the % purity of the copper?

1. Calculate the moles of thiosulfate used in the titration: $24.50 \times 10^{-3} \times 0.100 = 2.45 \times 10^{-3}$
2. Equation 3 shows that the number of moles of iodine is 0.5 times the moles of thiosulfate used. Moles $I_2 = 0.5 \times 2.45 \times 10^{-3} = 1.23 \times 10^{-3}$
3. Equation 2 shows that the number of moles of Cu^{2+} is 2 times the moles of iodine produced. Moles Cu^{2+} in the 25 cm^3 sample $= 2 \times 1.23$ $= 2.45 \times 10^{-3}$
4. The 25 cm^3 sample was removed from a total sample of 250 cm^3. Moles of Cu^{2+} in total sample $= 10 \times 2.45 \times 10^{-3}$ $= 2.45 \times 10^{-2}$
5. Equation 1 shows that 1 mole of Cu^{2+} comes from 1 mole Cu so the moles of Cu $= 2.45 \times 10^{-2}$ in the original impure copper sample.
6. A_r Cu = 63.5 Mass of 2.45×10^{-2} moles Cu = $2.45 \times 10^{-2} \times 63.5 = 1.56$ g
7. Percentage of Cu in the sample = $(1.56 \div 2.1) \times 100 = 74.3\%$

QUICK TEST

1. Write the electron configuration for Cr and for Cr^{3+}.
2. What is meant by a homogeneous catalyst?
3. Give an example of a heterogeneous catalyst and the reaction it catalyses.
4. How does having variable oxidation states help transition metals to act as catalysts?
5. What is meant by autocatalysis?
6. Name an oxidising agent that could be used in a redox titration.

PRACTICE QUESTIONS

1. $KMnO_4$ is a commonly used reagent in redox titrations. Which is true for a titration where $KMnO_4$ is in the burette? **[1 mark]**

   A $KMnO_4$ is the reducing agent. ☐

   B The colour change at the end point is pink to colourless. ☐

   C No indicator is used. ☐

   D The reaction mixture must be warmed. ☐

2. The electron configurations for Cu^{2+} and Fe^{3+} are: **[1 mark]**

	Cu^{2+}	Fe^{3+}	
A	Cu [Ar] $3d^7\ 4s^1$	Fe [Ar] $3d^4\ 4s^2$	☐
B	Cu [Ar] $3d^9$	Fe [Ar] $3d^6$	☐
C	Cu [Ar] $3d^9$	Fe [Ar] $3d^5$	☐
D	Cu [Ar] $3d^7\ 4s^1$	Fe [Ar] $3d^6\ 4s^2$	☐

3. About 80% of the rhodium produced is used as a heterogeneous catalyst in vehicle catalytic converters.

   a) What is meant by a heterogeneous catalyst? **[1 mark]**

   b) Describe the mechanism by which the catalyst works. **[4 marks]**

   c) Why are transition metals particularly suitable for this type of catalysis? **[1 mark]**

4. The reaction between iodide ions and peroxodisulfate ions can be catalysed by both Fe^{3+} and Fe^{2+}.

$$S_2O_8^{2-}(aq) + 2I^-(aq) \longrightarrow 2SO_4^{2-}(aq) + I_2(aq)$$

   a) Show how the reaction is catalysed by Fe^{3+}. **[3 marks]**

   b) How would the catalysis by Fe^{2+} differ from the method in part **a)** above? **[1 mark]**

   c) Explain why Zn^{2+} would not act as a catalyst for this reaction. **[2 marks]**

5. Brass is an alloy of copper and zinc. 0.500 grams of brass was dissolved in nitric acid and made up to 250 cm^3 with water.

   Equation 1: $Cu(s) \longrightarrow Cu^{2+} + 2e^-$

   25.0 cm^3 was removed and an excess of KI was added to it to convert the copper into copper iodide.

   Equation 2: $2Cu^{2+} + 4I^- \longrightarrow 2CuI + I_2$

   This sample was then titrated with 0.0150 mol dm^{-3} $NaS_2O_3(aq)$.

   Equation 3: $I_2 + 2S_2O_3^{2-} \longrightarrow 2I^- + S_4O_6^{2-}$

   The mean titre was 31.45 cm^3.

   Calculate the percentage of copper in the brass. **[5 marks]**

Complex Ions

A **complex ion** is a central metal ion surrounded by ligands. A **ligand** is a species that can donate a pair of electrons to form a coordinate bond with a central metal ion.

A **coordinate bond**, also known as a dative covalent bond, is where both electrons in the covalent bond come from the same atom. The **coordination number** of a complex ion is the number of coordinate bonds attached to the central metal ion.

Monodentate, also called **unidentate, ligands** can donate one pair of electrons to form one coordinate bond. Examples of monodentate ions:

H_2O NH_3 CN^- Cl^- OH^-

Bidentate ligands can donate two pairs of electrons to form two coordinate bonds with the central metal ion, e.g.

Ethanedioate Ethane-1,2-diamine

Multidentate ligands can form more than two coordinate bonds to a central metal ion, e.g. $EDTA^{4-}$ can form six bonds to a single ion, making it a hexadentate.

Shapes of Complex Ions

The shape of complex ions depends on the number of coordinate bonds. The charge is the sum of the charges on the ligand and the ion.

Coordination number 6 Octahedral	
monodentate ligands $[Cu(H_2O)_6]^{2+}$	bidentantate ligands $[Fe(C_2O_4)_3]^{3-}$

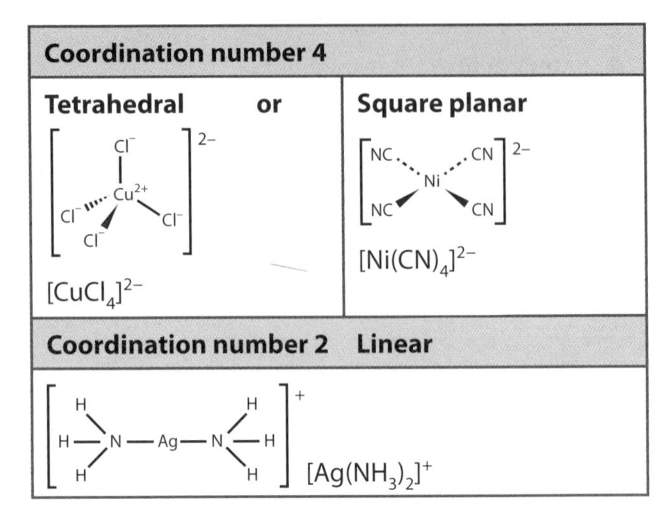

Coordination number 4	
Tetrahedral or	Square planar
$[CuCl_4]^{2-}$	$[Ni(CN)_4]^{2-}$
Coordination number 2 Linear	
$[Ag(NH_3)_2]^+$	

Stereoisomerism in Complex Ions

When there is more than one type of ligand around the central metal ion, the complex can have E/Z isomers.

E/Z or Cis-Trans Isomers

Cisplatin
cis or Z form

Transplatin
trans or E form

trans $[Cr(NH_3)_4Cl_2]^+$ cis $[Cr(NH_3)_4Cl_2]^+$

Optical Isomers

Octahedral complexes made from bidentate ligands have non-superimposable mirror images (see page 84).

Reactions of complex ions

Ligand Exchange/Substitution

A ligand in a complex may exchange for a different ligand. This typically results in a colour change and may change the coordination number, e.g.

$$[Cr(H_2O)_6]^{2+} + 6NH_3 \rightleftharpoons [Cr(NH_3)_6]^{2+} + 6H_2O(l)$$

red — purple

Change in colour but not in coordination number

$$[Co(H_2O)_6]^{2+} + 4Cl^- \rightleftharpoons [CoCl_4]^{2-} + 6H_2O(l)$$

pink — blue

Change in colour and change in coordination number from 6 to 4

Exchanging Monodentate for Multidentate Ligands

Bidentate ligands will exchange for monodentate ligands because the result is a more thermodynamically stable complex.

$$[Cu(H_2O)_6]^{2+}(aq) + 3C_2O_4^{2-} \rightleftharpoons [Cu(C_2O_4)_3]^{4-} + 6H_2O$$

The entropy change for this reaction, ΔS, is positive since there are 7 molecules in the products but only 4 in the reactants. Complexes with EDTA are very stable since substituting EDTA for any other ligand will give an increase in entropy and so Gibbs free energy is negative and the reaction is feasible.

$$[Cu(C_2O_4)_3]^{4-} + EDTA \rightleftharpoons [CuEDTA]^{4-} + 3C_2O_4^{2-}$$

The **chelate effect** is the name given to this increased stability of multidentate complexes.

Acid–Base Reactions of Complex Ions

In a complex ion where water is the only ligand, electrons from the oxygen of water molecules are donated to form the coordinate bond with the metal. This makes the O—H bond of the ligand more polar than it is in water. The water ligand can donate a proton to water, so the complex is an acid.

$$[Fe(H_2O)_6]^{2+} + H_2O \rightleftharpoons [Fe(H_2O)_5(OH)]^+ + H_3O^+$$

The higher the charge on the ion, the greater the polarisation of the O—H bond and the lower the pH of a solution of the ion. Ions with +3 oxidation number are more acidic than ions with +2 oxidation number.

When alkali is added to an aqua ion, an acid–base reaction occurs. e.g.

$$[Cu(H_2O)_6]^{2+}(aq) + 2OH^-(aq) \rightleftharpoons [Cu(H_2O)_4(OH)_2](s) + 2H_2O(l)$$

When $[Cu(H_2O)_6]^{2+}$ has donated two protons, the overall charge is 0 and $[Cu(H_2O)_4(OH)_2]$ precipitates as a light blue, gelatinous solid. Adding acid reverses the reaction and the precipitate re-dissolves to form the aqua ion.

Many 3+ ions first form an insoluble hydroxide with hydroxide ions and then re-dissolve as excess alkali is added. When all the H_2O ligands are replaced by OH^- the complex has a negative charge.

$$[Cr(H_2O)_6]^{3+}(\mathbf{aq}) + 3OH^-(aq) \longrightarrow [Cr(H_2O)_3(OH)_3](\mathbf{s}) + 3H_2O(l)$$

$$[Cr(H_2O)_3(OH)_3](\mathbf{s}) + 3OH^-(aq) \longrightarrow [Cr(OH)_6]^{3-}(\mathbf{aq}) + 3H_2O(l)$$

Why Complex Ions Are Coloured

Complexes with transition metals ions are coloured. The presence of ligands causes a change in the energy levels of the 3d orbitals in the central metal ion. Some have a decrease in energy and some an increase.

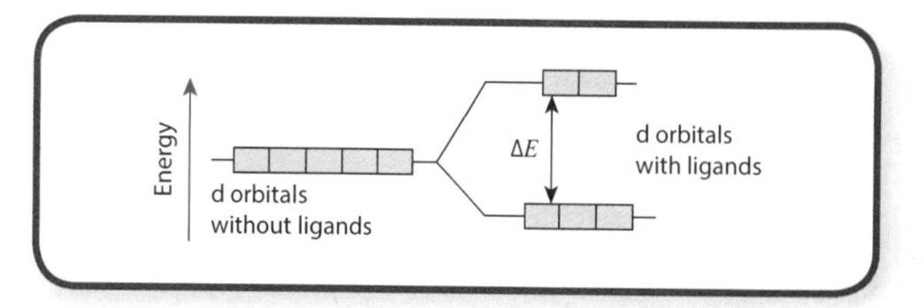

When white light strikes the molecule, electrons in the lower d orbitals are able to absorb a photon (packet of energy) of light and jump to orbitals at the higher level. The light that has not been absorbed is transmitted and this is the colour that is observed.

The relationship between the frequency and energy of light is $E = hf$ where h is Planck's constant and f is frequency.

Frequencies of light that are not absorbed are **transmitted** and seen as the **complementary colour**.

The colour absorbed, and so the colour seen, depends on the value of ΔE. Different metal ions and different ligands cause different energy gaps between the split d orbitals and so have different colours.

- $[Co(H_2O)_6]^{2+}$ is pink. The complex absorbs photons that give blue light. Blue light is high frequency so has high energy photons. ΔE is relatively large.
- $[CoCl_4]^{2-}$ is blue. It absorbs red light photons. These are of relatively low frequency. ΔE is relatively small.

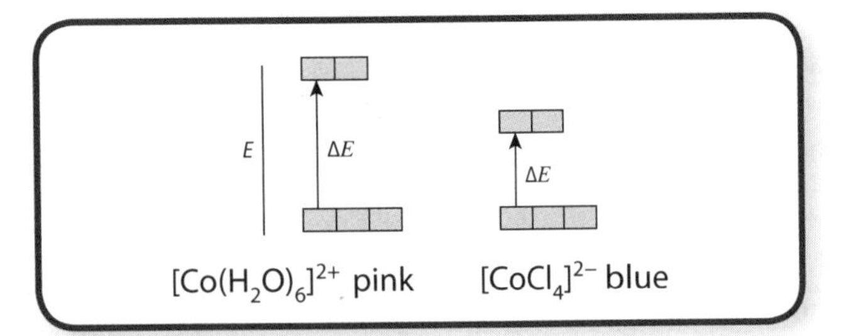

For the colours of complex ions see page 112.

Complex Ions in Use

Tollens' reagent also known as ammoniacal silver nitrate, contains the linear complex $[Ag(NH_3)_2]^+$.

Cisplatin is an anti-cancer drug (see page 90) with the formula $[PtCl_2(NH_3)_2]$ (see previous page). The cis configuration is essential to its function as a drug. Transplatin has no drug activity.

Haemoglobin contains a complex of Fe^{2+} with haem.

The haem molecule is planar allowing the central Fe^{2+} to accept two additional ligands above and below. The protein globin takes one space. The second position bonds to O_2 when the concentration is high, as in the lungs. In areas of low oxygen concentration, like the body tissues, a ligand exchange occurs. Oxygen is released and water bonds to the Fe^{2+}.

$[Fe(organic)O_2] + H_2O \rightleftharpoons [Fe(organic)H_2O] + O_2$

Carbon monoxide forms a strong bond to the central Fe in haem, which cannot be displaced by O_2. This means that once bonded to CO, haemoglobin is no longer able to function as an oxygen transporter.

Complexes and Redox Potential

(AQA only)

The value of $E^\ominus$ changes significantly when the water ligands in a metal complex are substituted.

The standard electrode potential is stated for a concentration of 1 mol dm^{-3}, e.g.

$Ni^{2+}(aq) + 2e^- \rightleftharpoons Ni(s) \quad E^\ominus = -0.25$ V

A ligand substitution reaction effectively decreases the concentration of the aqueous ion. This causes the position of equilibrium to move to the left.

$[Ni(NH_3)_6]^{2+}(aq) + 2e^- \rightleftharpoons Ni(s) + 6NH_3(aq) \quad E^\ominus = -0.51$ V

A change to the system that causes the position of equilibrium to move to the left, compared to standard conditions, results in a more negative $E^\ominus$.

QUICK TEST

1. What is the coordination number of the complex ion $[Cu(NH_3)_4(H_2O)_2]^{2+}$?
2. What is meant by a monodentate ligand?
3. Give the shape of $[CuCl_4]^{2-}$.
4. Why do multidentate ligands tend to exchange for monodentate ligands?
5. Is the following ligand exchange or acid–base?

   $[Fe(H_2O)_6]^{2+} + H_2O \rightleftharpoons [Fe(H_2O)_5(OH)]^+ + H_3O^+$
6. If red light is being absorbed what colour will be seen?

PRACTICE QUESTIONS

1. The complex ion $[Cu(NH_3)_4(H_2O)_2]^{2+}$ reacts with hydrochloric acid to produce a different complex, $[CuCl_4]^{2-}$. This results in a colour change from blue to yellow.

This reaction represents **[1 mark]**

	Type of reaction	Change in complex ion
A	acid–base	a change in coordination number
B	acid–base	no change in coordination number
C	ligand exchange	a change in coordination number
D	ligand exchange	no change in coordination number

2. Fe^{2+} and Fe^{3+} both form complex ions with water as the ligands. Which is **not** true? **[1 mark]**

A The shape of both complexes is octahedral.

B The two complexes have different colours.

C The two solutions have a different pH.

D Only Fe^{3+} will react with OH^-.

3. Fe^{3+} forms a complex ion with the formula $[Fe(H_2O)_6]^{3+}$. The colour of the solution formed is pale brown.

When aqueous thiocyanate ions are added to a solution of the complex the following reaction occurs:

$$[Fe(H_2O)_6]^{3+} + SCN^- \rightleftharpoons [Fe(H_2O)_5SCN]^{2+} + H_2O$$

The colour of the solution changes to deep red.

a) What **type** of reaction has occurred? **[1 mark]**

b) How has the coordination number of the complex changed? **[1 mark]**

c) Draw the structure of the $[Fe(H_2O)_6]^{3+}$ ion and name its shape. **[3 marks]**

d) Explain why the $[Fe(H_2O)_6]^{3+}$ ion is coloured and why it changes colour on addition of SCN^-. **[6 marks]**

4. Ethandioate and ethyldiamine are both bidentate ligands that will form complex ions with Cr^{3+}.

a) Explain what is meant by a bidentate ligand. **[1 mark]**

b) Draw the structure of the complex that Cr^{3+} makes with ethyldiamine. **[2 marks]**

The $[Cu(H_2O)_6]^{2+}$ ion reacts with two molecules of ethandioate in a partial ligand exchange to form $[Cu(C_2O_4)_2(H_2O)_2]^{2-}$.

c) Explain why there are two isomers of the ion that is formed. Use a diagram to illustrate your answer. **[4 marks]**

When EDTA is added to the complex a further ligand exchange occurs and all ligands other than EDTA are displaced.

d) Write an equation for this reaction. **[1 mark]**

e) Why does EDTA displace all other ligands? **[1 mark]**

Period 3 Elements and Their Oxides

Period 3 elements from Na to S all react directly with oxygen.

				mpt of oxide/°C
Na	$4Na(s) + O_2(g)$	$\longrightarrow$	$2Na_2O(s)$	1548
Mg	$2Mg(s) + O_2(g)$	$\longrightarrow$	$2MgO(s)$	3125
Al	$4Al(s) + 3O_2(g)$	$\longrightarrow$	$2Al_2O_3(s)$	2345
Si	$Si(s) + O_2(g)$	$\longrightarrow$	$SiO_2(s)$	1883
P	$P_4(s) + 5O_2(g)$	$\longrightarrow$	$P_4O_{10}(s)$	853
S	$S(s) + O_2(g)$	$\longrightarrow$	$SO_2(g)$	200

S also forms the oxide SO_3 when the reaction is catalysed (see page 57).

The melting points of Period 3 oxides reflect their bonding and structure. Na $\longrightarrow$ Al are ionic; Si is giant covalent; P and S are simple covalent.

Reactions of Period 3 Oxides with Water

Na and **Mg** oxides react with water to form alkaline solutions.

$Na_2O + H_2O \longrightarrow 2NaOH$
NaOH is very soluble and gives a high pH solution.

$MgO + H_2O \longrightarrow Mg(OH)_2$
$Mg(OH)_2$ is only slightly soluble so gives a lower pH solution than NaOH.

Al and **Si** oxides are insoluble and do not react with water.

P and S Oxides React With Water

$P_4O_{10} + 6H_2O \longrightarrow 4H_3PO_4$
H_3PO_4 is a weak tribasic acid with a higher pH than strong acids.

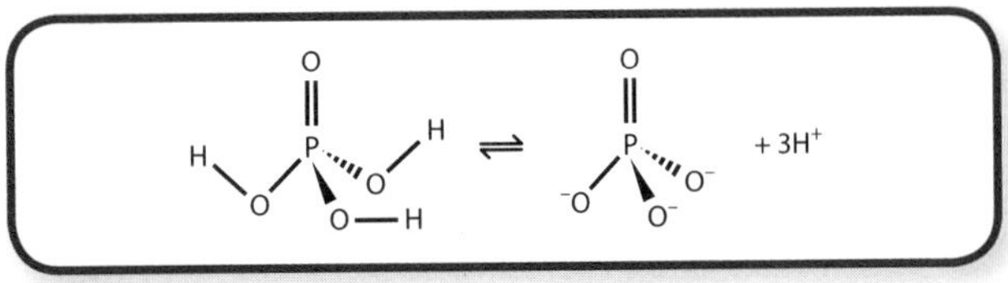

All three of the OH groups can dissociate; each has its own pK_a.

$SO_2 + H_2O \rightleftharpoons H^+ + HSO_3^-$
SO_2 theoretically produces H_2SO_3 but this only exists in ionised form.

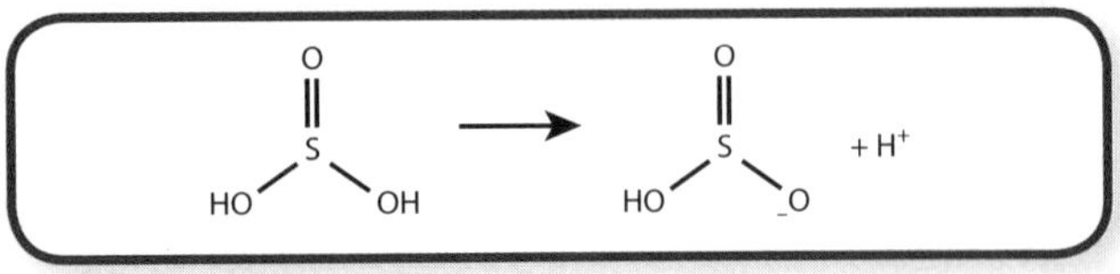

$SO_3 + H_2O \longrightarrow H_2SO_4$
H_2SO_4 is a dibasic, strong acid giving very low pH solutions. The second dissociation is weaker than the first.

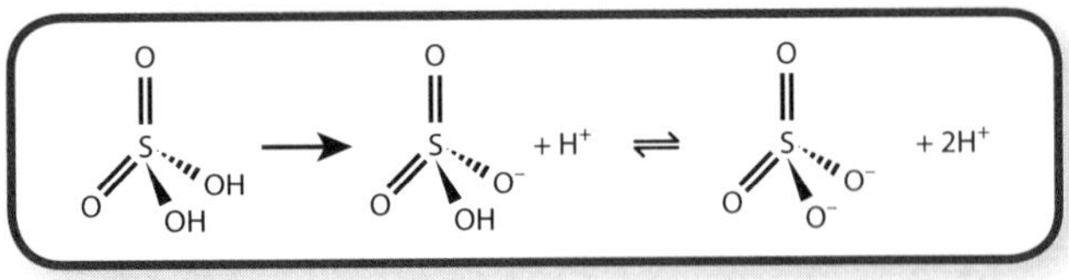

Reactions of Period 3 Oxides with Acids and Bases

The basic oxides react with acids to give a salt + water, e.g.

$$Na_2O + 2HCl \longrightarrow 2NaCl + H_2O$$

$$MgO + H_2SO_4 \longrightarrow MgSO_4 + H_2O$$

Aluminium oxide is amphoteric and will react with both acids and bases.

$$Al_2O_3 + 6H^+ \longrightarrow 2Al^{3+} + 3H_2O$$

$$Al_2O_3 + 3H_2SO_4 \longrightarrow Al_2(SO_4)_3 + 3H_2O$$

Aluminium oxide reacts with bases to form aluminates, negatively charged compound ions.

$$Al_2O_3 + 2OH^- + 3H_2O \longrightarrow 2Al(OH)_4^-$$

$$Al_2O_3 + 2NaOH + 3H_2O \longrightarrow 2NaAl(OH)_4$$

SiO_2 is acidic because it reacts with alkali. It does not react with aqueous alkali because it is insoluble but will react with molten sodium hydroxide.

$$SiO_2(s) + 2NaOH(l) \longrightarrow Na_2SiO_3(l) + H_2O(g)$$

Phosphorus and sulfur oxides react with the water in aqueous alkali to form an acid, which then reacts with the alkali to form a salt + water.

$$H_3PO_4 + 3NaOH \longrightarrow Na_3PO_4 + 3H_2O$$

Since the H_3PO_4 can dissociate to form $H_2PO_4^-$ and HPO_4^{2-}, the salt formed will depend on the ratio of base to acid.

Sulfur oxides react in a similar way.

$$H_2SO_4 + 2NaOH \longrightarrow Na_2SO_4 + 2H_2O$$

p-block Chemistry

p-block elements have their highest energy electron in a p orbital.

The metal/non-metal divide passes diagonally thorough the block.

Group					
3	4	5	6	7	8
B	**C**	**N**	**O**	**F**	**Ne**
Al	**Si**	**P**	**S**	**Cl**	**Ar**
Ga	**Ge**	**As**	**Se**	**Br**	**Kr**
In	**Sn**	**Sb**	**Te**	**I**	**Xe**
Tl	**Pb**	**Bi**	**Po**	**At**	**Rn**

p-block Elements Do Not Always Obey the Stable Octet Rules

Some Group 3 elements are able to form stable covalent compounds with only 6 electrons in their outer shell, e.g. $AlCl_3$ and BF_3.

$AlCl_3$ dimerises sharing two dative covalent bonds so that it now obeys the octet rule.

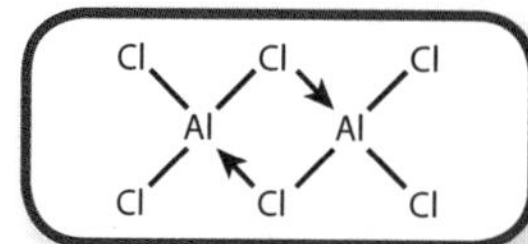

The radius of the B atom is too small to allow BF_3 to dimerise but it is able to accept a pair of electrons from molecules such as NH_3 to form **donor–acceptor compounds**.

$$H_3N: + BF_3 \longrightarrow H_3N \rightarrow BF_3$$

Some Group 5, 6 and 7 elements are able to form stable covalent compounds with more than 8 electrons in their outer shell, e.g. PCl_5 and SF_6.

This is because the lone pair electrons are promoted into the empty d orbitals to make one electron per orbital. Each electron can then form a covalent bond.

Boron Nitride Forms Giant Covalent Structures

Hexagonal boron nitride has a similar structure to graphite. It is soft and used as a lubricant because the layers are able to slide over each other. Cubic boron nitride has a similar structure to diamond. It is hard and used as an abrasive.

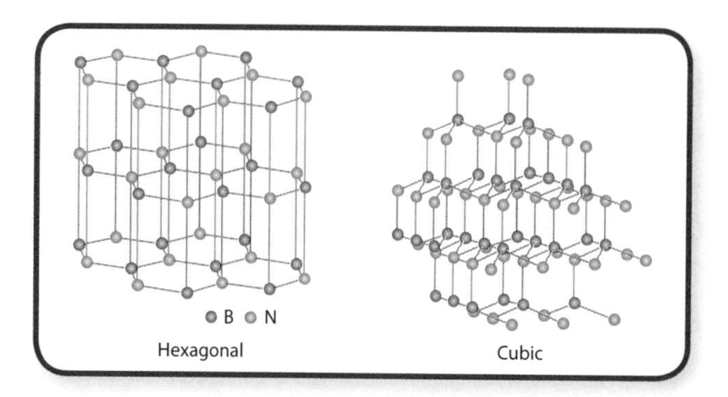

Some p-block Elements Are Amphoteric

Some p-block elements can react with both acids and bases, e.g. aluminium.

Acid: $2Al + 6HCl \longrightarrow 2AlCl_3 + 3H_2$

Alkali: $2Al + 2NaOH + 6H_2O \longrightarrow 2NaAl(OH)_4 + 3H_2$

Both $Al^{3+}(aq)$ and $Pb^{2+}(aq)$ form a precipitate when hydroxide ions are added. The precipitate re-dissolves on addition of more hydroxide. This process can be reversed by addition of hydrogen ions.

$$Pb^{2+}(aq) \underset{H^+}{\overset{OH^-}{\rightleftharpoons}} Pb(OH)_2(s) \underset{H^+}{\overset{OH^-}{\rightleftharpoons}} [Pb(OH)_4]^{2-}(aq)$$

In Groups 3–5 Stability of the Lower Oxidation State Increases Down the Group

p-block elements have electrons in the s and p orbitals of their outer shell. When bonding they have the potential to use either only p orbital electrons (the lower oxidation state) or both p and s orbital electrons (the higher oxidation state), e.g. SnO and SnO_2.

Going down Groups 3, 4 and 5 there is increasing tendency not to use the s orbital electrons. This is known as the inert pair effect.

Top of Group 4: Carbon in the +4 oxidation state is more stable than in the +2 state. This makes CO a good reducing agent, being readily oxidised to CO_2.

$$Fe_2O_3 + 3CO \longrightarrow 2Fe + 3CO_2$$

Bottom of Group 4: Lead in the +2 state is more stable than in the +4 state. This makes PbO_2 a good oxidising agent, being readily reduced to PbO.

$$PbO_2 + 4HCl \longrightarrow PbCl_2 + Cl_2 + 2H_2O$$

Metallic Character Increases Down the Group

The metal/non-metal divide is crossed moving down Groups 3–6. This is reflected in the properties of the oxides and chlorides.

At the top of Group 4, CO_2 is a simple covalent compound with a low boiling point. It is an acidic oxide, reacting with bases.

$$CO_2 + Ca(OH)_2 \longrightarrow CaCO_3 + H_2O$$

At the bottom of Group 4, PbO has a giant ionic structure with a high melting point. It exists in two crystal forms (allotropes), one red and one yellow. It is amphoteric.

Some Properties of Chlorides of Group 4

Compound	Bonding	Reaction with H_2O
CCl_4	covalent	no reaction
$SiCl_4$	covalent	$SiCl_4 + 4H_2O \longrightarrow SiO_2 + 4HCl$
$PbCl_2$	ionic	dissolves in hot water

Some Reactions of Pb^{2+}

$Pb^{2+}(aq) + 2Cl^-(aq) \longrightarrow PbCl_2(s)$ a white precipitate

$Pb^{2+}(aq) + 2I^-(aq) \longrightarrow PbI_2(s)$ a yellow precipitate

See *Letts A-level Chemistry Year 1 (and AS) in a Week* for more information on reactions of Cl_2 and halides.

QUICK TEST

1. Write an equation for the reaction of phosphorus with oxygen.
2. What is the reaction of SiO_2 with water?
3. Give the equation for the overall reaction of $SO_3(g)$ with NaOH(aq).
4. What is a donor–acceptor molecule?
5. What is the inert pair effect?
6. Write the equation for the reaction of $SiCl_4$ with water.

PRACTICE QUESTIONS

1. Magnesium oxide is **[1 mark]**

   A a neutral oxide ☐ **B** an acidic oxide ☐

   C a basic oxide ☐ **D** an amphoteric oxide. ☐

2. The melting point of phosphorus oxide is **[1 mark]**

   A lower than SO_2 ☐ **B** higher than Al_2O_3 ☐

   C lower than Na_2O ☐ **D** higher than MgO. ☐

3. Group 4 elements all have the same number of electrons in their outer shell but do not all react in the same way.

   a) Write the electron structure for germanium. **[1 mark]**

   Carbon and lead both form compounds with oxidation state +2 and +4.

   b) Write formulae for a carbon compound and a lead compound in the +4 oxidation state. **[2 marks]**

   The most stable lead compound exists in the +2 oxidation state.

   c) Explain how it is possible for lead to form compounds in the +2 oxidation state. **[2 marks]**

   d) Comment on the stability of carbon compounds in the +2 oxidation state. **[1 mark]**

4. Aluminium and aluminium oxide can both react with both acids and bases.

   a) What name is given to compounds that react with both acids and bases? **[1 mark]**

   b) Write equations showing the reaction of aluminium oxide with sodium hydroxide and with hydrochloric acid. **[2 marks]**

   Aluminium chloride is a compound of a metal and a non-metal with some surprising properties. It does not conduct electricity as a solid or when molten. At 1 atm it sublimes at below 200°C.

   c) Draw the structure of aluminium chloride vapour. **[1 mark]**

5. Boron nitride is an extremely stable, chemically unreactive compound that has uses as an abrasive and as a lubricant.

   a) Draw a dot and cross diagram for a boron atom and a nitrogen atom. **[1 mark]**

   b) Explain how nitrogen is able to form four bonds with boron. **[1 mark]**

   c) Explain how boron and nitrogen can form a compound that can act as both a lubricant and an abrasive. You should describe structures in your answer. You may include diagrams. **[5 marks]**

Carbonyl Compounds

Aldehydes and ketones both contain a carbonyl group.

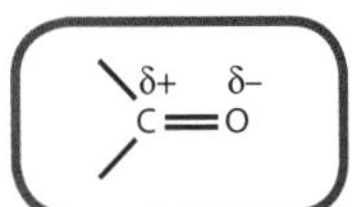

Testing for Carbonyl Groups with 2,4-DNP

A solution of 2,4-dinitrophenylhydrazine reacts with a carbonyl group to give a yellow-orange precipitate.

Identification of Carbonyl Compounds with 2,4-DNP

The crystals formed have very sharp melting points. This can be compared against a database to identify the original carbonyl compound from which they formed. The crystals should be purified by re-crystallisation before the melting point is measured.

Solubility of Carbonyl Compounds

Carbonyl compounds contain a polar covalent bond and so can form dipole–dipole bonds with each other. They have no H covalently bonded to O, N or F so cannot hydrogen bond with each other. This gives them a lower boiling point than alcohols of similar M_r. The lone pairs of electrons on oxygen means that carbonyl compounds are able to hydrogen bond with water.

Aldehydes

Functional group:

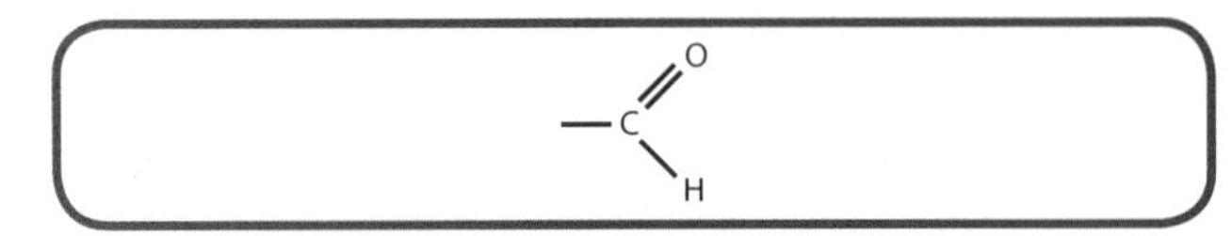

The aldehyde group is always at the end of a carbon chain.

The formula is written **R—CHO** not R—COH to distinguish aldehydes from alcohols.

There is an H attached to the carbon of the carbonyl group. This H makes aldehydes good reducing agents.

Aldehyde Names End in -al

They are named in the same way as alkanes but with the ending -al.

Ethanal

3-methylbutanal

Aldehydes Can Be Made By Oxidation of Primary Alcohols

Primary alcohols can be oxidised to aldehydes by heating with acidified potassium dichromate. The reaction mixture is distilled to remove the aldehyde as it is made and prevent further oxidation to a carboxylic acid.

[O] represents the oxidising agent.

Ethanol → (2[O]) Ethanal + $2H_2O$

Oxidation to Carboxylic Acids

Aldehydes can be oxidised to carboxylic acids and this is the basis of the test for aldehydes. Heating with acidified potassium dichromate gives a colour change from orange to green.

Fehling's Solution Gives a Brick-Red Precipitate with an Aldehyde

Fehling's solution contains copper ions and potassium hydroxide. The Cu^{2+} is reduced as the aldehyde is oxidised.

$$2Cu^{2+}(aq) + 2e^- + 2OH^-(aq) \longrightarrow Cu_2O(s) + H_2O(l)$$

Tollens' Reagent Gives a Silver Mirror with Aldehydes

Tollens' reagent contains silver nitrate and ammonia. The silver ions are reduced by the aldehyde.

$$Ag^+(aq) + e^- \longrightarrow Ag(s)$$

Aldehydes Can Be Reduced to Primary Alcohols with $NaBH_4$(aq)

The reducing reagent can be considered to be H^-.

$$R{-}\overset{\delta+}{C}(H){=}\overset{\delta-}{O} + :H^- \longrightarrow R{-}C(H)(H){-}O^- \;\; H{-}O{-}H \longrightarrow R{-}C(H)(H){-}OH + OH^-$$

An alternative reagent is $LiAlH_4$ in dry ether.

Overall equation:

$$CH_3CHO + 2[H] \longrightarrow CH_3CH_2OH$$

The reducing agent is represented as [H].

Ketones

Functional group:

$$R(R')C{=}O$$

There is no H on the carbon of the carbonyl group. This means that they are not good reducing agents.

Ketone Names End in -one

Ketones are named as alkanes but with the ending -one. The position of the carbonyl group is indicated by a number immediately before the -one or in front of the name.

$$CH_3{-}CH(CH_3){-}CH_2{-}C({=}O){-}CH_3$$

4-methylpentan-2-one

Ketones Can Be Made By Oxidation of Secondary Alcohols

Heating a secondary alcohol with acidified potassium dichromate produces a ketone. There is no need to distil the product off as it forms since the ketone cannot be oxidised further by the dichromate.

$$R'{-}CH(OH){-}R \xrightarrow{H^+/Cr_2O_7^{2-}} R'{-}C({=}O){-}R$$

Ketones are reduced to secondary alcohols by $NaBH_4$ or $LiAlH_4$.

Carbonyl Groups Undergo Nucleophilic Addition with Cyanide Ions to Form Hydroxynitriles

This is done using KCN followed by dilute acid at room temperature.

The lone pair of electrons on the **C** of the cyanide ion is donated to the δ+ carbon of the carbonyl group.

$$R{-}\overset{\delta+}{C}(H){=}\overset{\delta-}{O} + :CN^- \longrightarrow R{-}C(H)(CN){-}O^- \;\; H^+ \longrightarrow R{-}C(H)(CN){-}OH$$

Alternative reagents: HCN with KCN; HCN with a small amount of KOH. All of these provide CN^- for the first step of the reaction and H^+ for the second step.

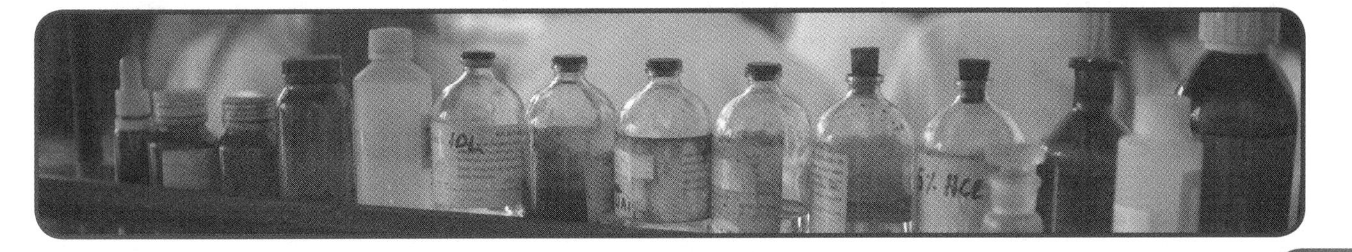

A hydroxynitrile is also known as a cyanohydrin.

This reaction increases the carbon chain length by 1.

propanone ⟶ 1-hydroxy**butane**nitrile

Nucleophilic addition of HCN to carbonyl compounds may result in the production of optical isomers (see page 84).

OH, C, R, H, CN | OH, C, H, R, CN

The carbon of the carbonyl becomes a chiral centre after the addition of the HCN. The bond angles around the carbonyl carbon are approximately 120°, giving a planar shape. This means that it is equally likely that the CN^- will attack either side of the molecule resulting in a racemic mixture (see page 85). The production of a racemic mixture supports the proposed reaction mechanism.

For ketones with two identical R groups, no chiral centre is produced, e.g. propanone.

$$CH_3COCH_3 \xrightarrow[H^+]{CN^-} CH_3C(OH)(CN)CH_3$$

The nitrile group is easily oxidised to a carboxylic acid by warming with HCl(aq).

$$RCH(OH)CN \xrightarrow{HCl(aq)} RCH(OH)COOH + NH_4Cl$$

The Iodoform Test

The iodoform test identifies CH_3CO groups. A carbonyl group that is adjacent to a methyl group gives a yellow precipitate of CHI_3 (iodoform) when warmed with I_2(aq) and KOH(aq).

$$RCOCH_3 \xrightarrow{OH^-/I_2} RCOO^- + CH_3I$$

The only aldehyde that contains CH_3CO is ethanal. All 2-ketones (ketones with the carbonyl on carbon 2) contain this group.

Ethanol and all 2-alcohols react in the same way.

QUICK TEST

1. Draw the structure of pentan-3-one.
2. Outline a test for the presence of a carbonyl group.
3. Why is there not a ketone with the name butan-1-one?
4. Name a reagent that could be used to distinguish propanone from propanal.
5. What reagent would you use to reduce an aldehyde?
6. Suggest two reaction steps to convert an aldehyde into a carboxylic acid.

PRACTICE QUESTIONS

1. Which statement is incorrect? **[1 mark]**

   A Carbonyl compounds contain polar bonds. ☐

   B Carbonyl compounds can hydrogen bond to water. ☐

   C Carbonyl compounds can hydrogen bond to each other. ☐

   D Carbonyl compounds undergo electrophilic addition. ☐

2. Which statement is true of propanal? **[1 mark]**

   A It gives a yellow precipitate when warmed with sodium hydroxide and iodine. ☐

   B It gives a silver mirror with ammoniacal silver nitrate. ☐

   C The product of its reaction with HCN is not a racemic mixture. ☐

   D It can be reduced to a secondary alcohol with $NaBH_4$. ☐

3. A chemist has isolated a compound with a suspected carbonyl group.

   a) What test tube test could be performed to confirm that a carbonyl group is present? **[2 marks]**

   b) How could the chemist decide if the carbonyl group was part of an aldehyde or a ketone? **[2 marks]**

   c) Suggest a way in which the chemist might be able to identify which aldehyde or ketone was present. **[2 marks]**

4. This question is about butanone.

   a) Draw the structure of butanone and indicate any polar covalent bonds by showing partial charges. **[2 marks]**

   b) Butanone can be formed from an alcohol. Give the name of the alcohol and the reagents and conditions required. **[3 marks]**

   The reaction can be reversed by reduction with $NaBH_4$.

   c) Draw the mechanism for the reduction. Use H^- to represent the reducing agent. **[3 marks]**

   Butanone can be converted into a carboxylic acid in a two-step procedure. The first step involves mixing the butanone with KCN followed by dilute acid.

   d) Draw the mechanism for the reaction of butanone with KCN followed by dilute acid. **[4 marks]**

   The product of this reaction is a racemic mixture.

   e) Explain why the reaction produced optical isomers and why a racemic mixture is made. **[3 marks]**

   f) What further step must be followed to convert the product in part **d)** to a carboxylic acid? **[1 mark]**

   g) Name the carboxylic acid that would be made. **[1 mark]**

Carboxylic Acids and Their Derivatives

Carboxylic acids contain the functional group —COOH.

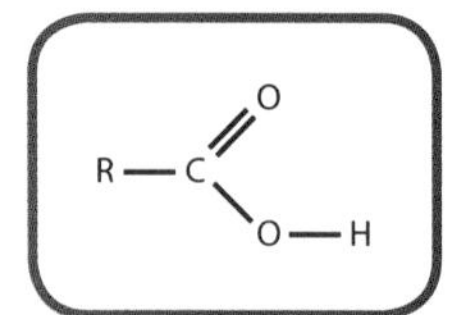

They are named as alkanes but with the ending -oic acid.

The functional group is always at the end of the carbon chain and makes the carbon of the COOH carbon number 1 in the chain.

If there are 2 —COOH groups at either end of the carbon chain then the molecule is a dioic acid.

3-methylpentandioic acid

If a COOH is present as a side group then the prefix carboxy- is used. The carbon atom to which it is attached must be numbered.

3-carboxy-3-hydroxypentandioic acid

Carboxylic Acids Can Hydrogen Bond

Hydrogen bonding with water means that short chain carboxylic acids are water soluble. Hydrogen bonding with themselves gives carboxylic acids a high boiling point compared to similar M_r hydrocarbons.

The bond angle around the —COOH allows two molecules to form a dimer held together with two hydrogen bonds.

Carboxylic acid dimer

Gaseous carboxylic acids tend to form dimers.

Carboxylic Acids Only Partially Dissociate

$$R\text{—}COOH \rightleftharpoons R\text{—}COO^- + H^+$$

Carboxylic acids react as typical acids:

- with a base to produce a salt + water

$$CH_3COOH + NaOH \longrightarrow CH_3COONa + H_2O$$

- with a reactive metal to produce a salt + hydrogen

$$2CH_3COOH + Mg \longrightarrow (CH_3COO)_2Mg + H_2$$

- with a carbonate to produce a salt + carbon dioxide + water

$$2CH_3COOH + Na_2CO_3 \longrightarrow 2CH_3COONa + H_2O + CO_2$$

Forming Carboxylic Acids

Alcohol to carboxylic acid: Oxidation of primary alcohols or aldehydes with acidified potassium dichromate results in carboxylic acids (see page 68).

Nitriles to carboxylic acid by refluxing with dilute acid.

$$R\text{—}CN + HCl + 2H_2O \longrightarrow RCOOH + NH_4Cl$$

Nitriles can also be hydrolysed with alkali giving the salt of the carboxylic acid.

Aromatic compound to carboxylic acid: A benzene ring with any alkyl side chain can be oxidised to benzoic acid by heating with alkaline potassium manganate.

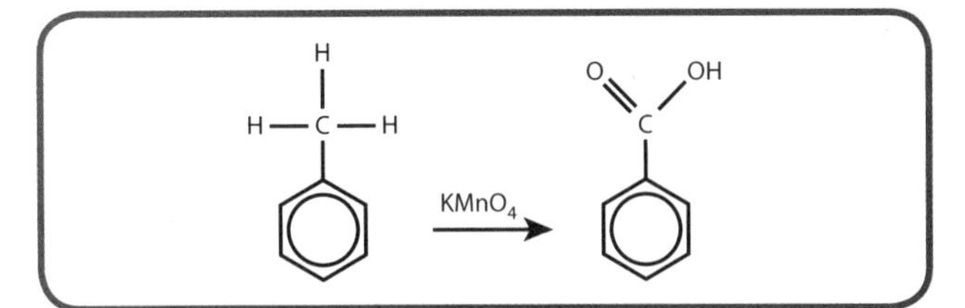

The result of the oxidation reaction is benzoic acid, regardless of the length of the alkyl chain.

Reactions of Carboxylic Acids

Carboxylic acid to primary alcohol by reduction with $LiAlH_4$ in dry ethoxyethane followed by dilute sulfuric acid. ([H] represents the reducing agent.)

The reduction does not occur with $NaBH_4$.

Carboxylic acid to alkane (decarboxylation): The carboxylic acid group can be removed to give an alkane with one less carbon atom. The reagent is soda lime, which is a mixture of sodium hydroxide and calcium oxide that is added to the sodium salt of the carboxylic acid.

$$CH_3COONa + NaOH \longrightarrow CH_4 + Na_2CO_3$$

Benzoic acid is also decarboxylated in this way to give benzene.

$$C_6H_5COOH + NaOH \longrightarrow C_6H_6 + Na_2CO_3$$

Carboxylic acid to acyl chlorides:

The reaction produces 'steamy fumes' as the HCl dissolves into water vapour in the air, making droplets of hydrochloric acid. The acyl chloride can be separated from the $POCl_3$ by fractional distillation.

Acyl chlorides react in a similar way to carboxylic acids but more vigorously. They cannot be used in solution.

Carboxylic acid to ester: Heat with an alcohol in the presence of a concentrated acid catalyst.

Esters

An ester contains the group:

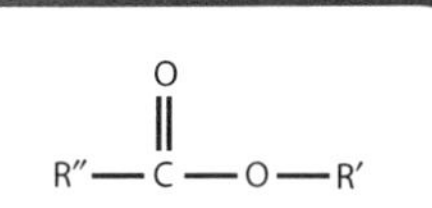

To name an ester:

1. Name the R group directly attached to the O (labelled R′ above). Add the ending -yl.
2. Name the other R group attached to the carbonyl group (labelled R″ above) with the ending -oate.

Methylpropanoate

Esters can be made by reaction of alcohols with carboxylic acids, acyl chlorides or acid anhydrides.

These all contain the same group, known as an acyl group but eliminate a different molecule.

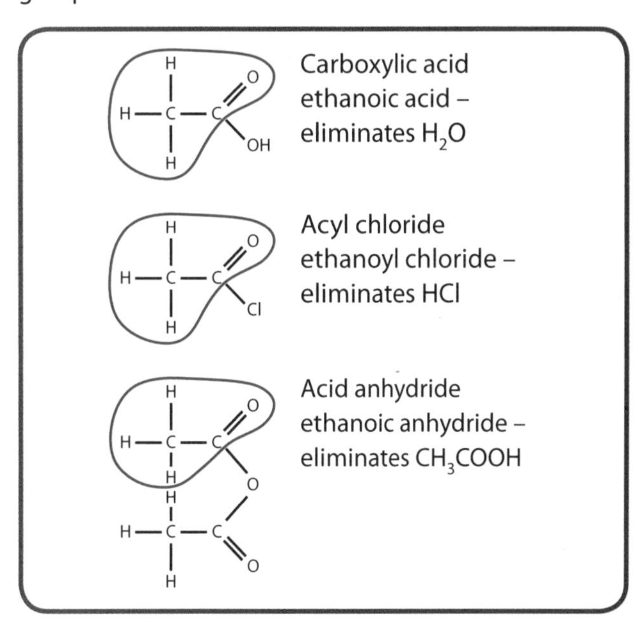

Hydrolysis of Esters

Esters can be hydrolysed with either acid or alkali (moderately concentrated).

$$CH_3COOCH_3 + H_2O \xrightarrow{H^+} CH_3COOH + CH_3OH$$

$$CH_3COOCH_3 + H_2O \xrightarrow{OH^-} CH_3COO^- + CH_3OH$$

The –OH in Carboxylic Acids Is More Acidic Than in Alcohols and Phenols

The conjugate base of a carboxylic acid is stabilised by delocalisation of the electrons across the COO group. The charge is shared between the two oxygens and the bond length of both C—O bonds is halfway between a double and single bond.

$$R{-}COOH \rightleftharpoons RCOO^- + H^+$$

R—C(delocalised O, O)⁻ not R—C(=O)O⁻

This makes the O—H bond in COOH more acidic than when it is found in other environments, such as in alcohols or water.

$$RCH_2OH \rightleftharpoons RCH_2O^- + H^+ \qquad H{-}O{-}H \rightleftharpoons OH^- + H^+$$

The conjugate base of an alcohol is very unstable as the negative charge is centred on the oxygen and the equilibrium is far to the left. Neither water nor alcohols have a pH below 7. They do not take part in typical acid reactions.

The —OH group in phenol is slightly acidic.

$$C_6H_5{-}O{-}H \rightleftharpoons C_6H_5{-}O^- + H^+$$

The conjugate base is stabilised by some delocalisation of electrons into the delocalised ring of electrons in benzene. Position of equilibrium is further to the right than water but not as far as carboxylic acids.

Phenol is acidic enough to react with a base to form a salt but not acidic enough to react with a carbonate to form carbon dioxide and water.

$$C_6H_5{-}O{-}H + NaOH \rightarrow C_6H_5{-}ONa + H_2O$$

The stability of the conjugate base is reflected in the pH of these compounds and in their reactions.

QUICK TEST

1. Draw the skeletal structure of butanedioic acid.
2. Write an equation for the reaction of Ca with ethanoic acid.
3. How could you convert ethanoic acid to ethanol?
4. What is eliminated when ethanoyl chloride and butanol form an ester?
5. What is the name of the ester formed?
6. Why does a carboxylic acid have a lower pH than an alcohol?

PRACTICE QUESTIONS

1. Lactic acid has the following structure:

$CH_3CH(OH)COOH$

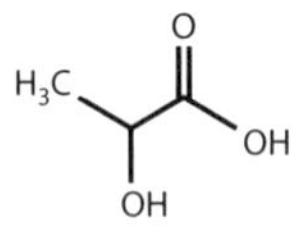

The systematic name for lactic acid is **[1 mark]**

A hydroxyethanoic acid ☐

B ethandioic acid ☐

C 2-methyl-2-hydroxyethanoic acid ☐

D 2-hydroxypropanoic acid ☐

2. Which is true of carboxylic acids? **[1 mark]**

A They are less reactive than acyl chlorides. ☐

B They cannot be used in aqueous solution. ☐

C They react with acidified potassium dichromate to form ketones. ☐

D They are made from amines by refluxing with dilute acid. ☐

3. Hexanedioylchloride is a useful starting material in the production of polymers.

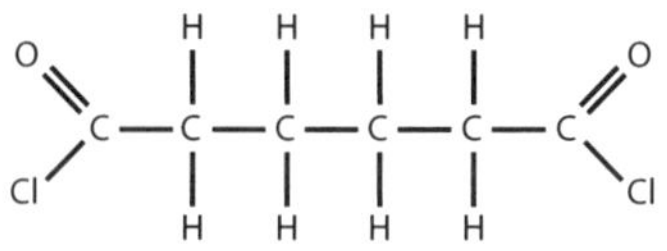

$ClOC-CH_2-CH_2-CH_2-CH_2-COCl$

a) Suggest a way that hexanedioylchloride could be made from hexane-1,6-diol. **[4 marks]**

b) Draw the structure of the organic product if hexanedioylchloride was reacted with ethanol. **[2 marks]**

c) Name the functional group in this product. **[1 mark]**

d) What other substance is produced as a result of this reaction? **[1 mark]**

4. The following structures all contain a C—O—H group:

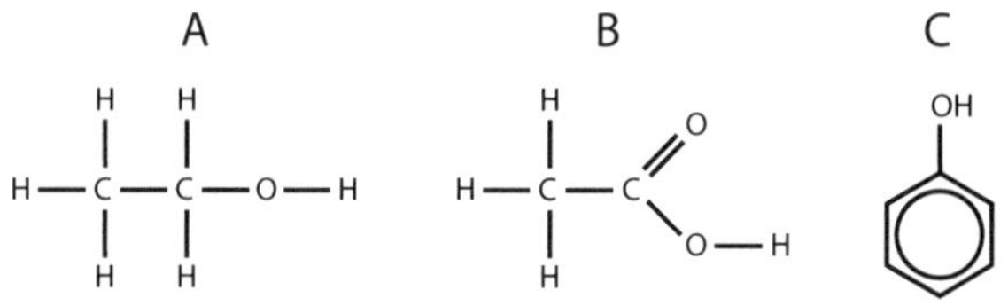

A: $H-CH_2-CH_2-O-H$ B: $H-CH_2-C(=O)-O-H$ C: phenol (C_6H_5OH)

a) Identify the functional group in each molecule. **[3 marks]**

b) Compare the reaction of A and C with NaOH. **[2 marks]**

c) Compare the reaction of B and C with $NaHCO_3$. **[2 marks]**

d) Use your knowledge of acids and equilibria to explain the difference in behaviour of the OH group in these three molecules. **[4 marks]**

Benzene

Benzene

Molecules containing a benzene ring are classed as **aromatic**. Aromatic hydrocarbons are called arenes.

Benzene with one hydrogen missing is known as a phenyl group C_6H_5–.

Where benzene is the parent functional group the name of the compound ends in -benzene. Where it is not the parent functional group it is described by the prefix phenyl-.

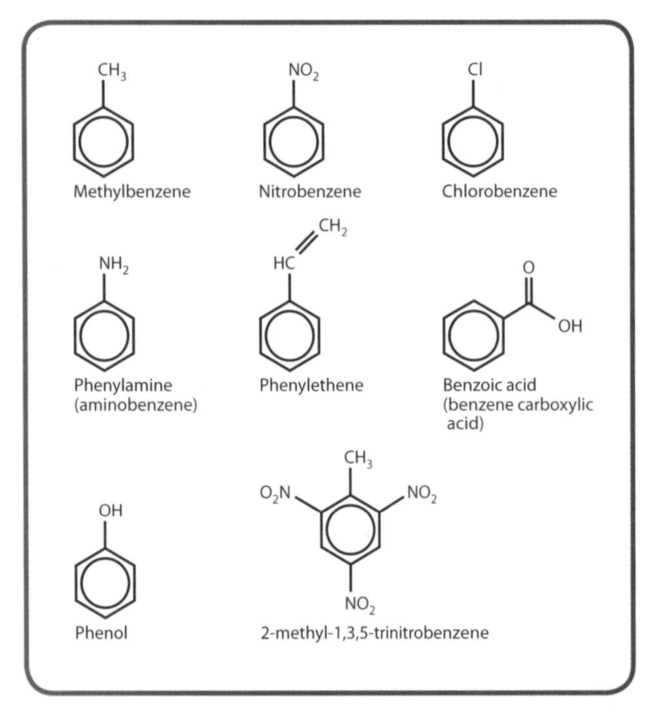

The carbon to which a group is attached becomes carbon 1.

The Structure of Benzene

Each carbon in the 6-carbon ring forms two σ bonds to the two carbons on either side and one σ bond to a hydrogen atom. Three regions of electron density around each carbon give a bond angle of 120° so the benzene ring is planar.

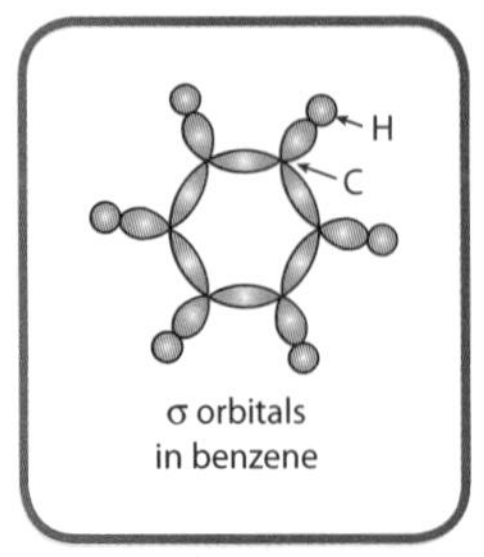

σ orbitals in benzene

This leaves one electron per carbon atom in a p orbital at right angles to the three σ bonds. The p orbitals overlap to form a ring of delocalised electrons above and below the plane of the ring, which gives the molecule an increased stability.

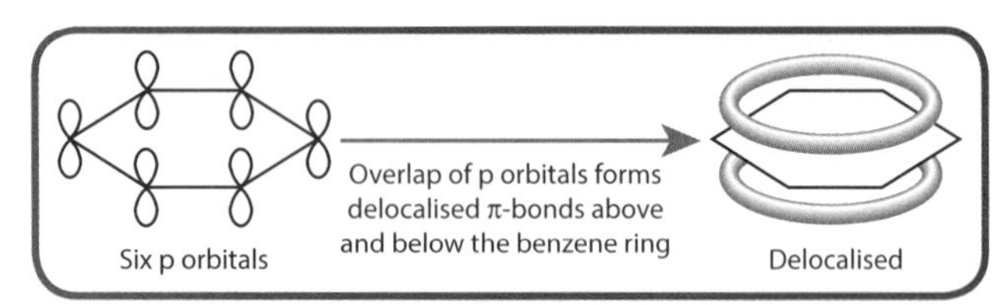

Evidence Against the Kekulé Structure of Benzene

Kekulé proposed a structure for benzene that is cyclohexa-1,3,5-triene.

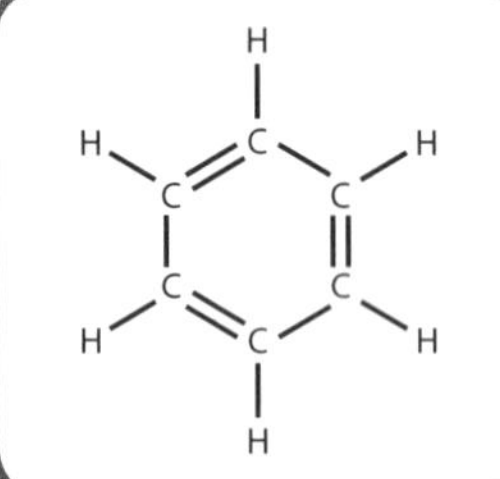

Later evidence showed that this structure cannot be correct because:

- all the carbon–carbon bonds in benzene are of the same length but C—C single bonds are longer than C═C double bonds. The use of x-ray crystallography showed that the bond length is an average of the two and could be considered as a 1.5 bond between carbon atoms
- benzene does not undergo electrophilic addition, a typical alkene reaction, e.g. it does not decolourise bromine water
- the enthalpy of hydrogenation of benzene is much smaller than the expected value.

+ $3H_2$ → $\Delta H^{\ominus}$ −208 KJ mol^{-1}

+ H_2 → $\Delta H^{\ominus}$ −120 KJ mol^{-1}

" " + $3H_2$ → Predicted $\Delta H^{\ominus}$ −360 KJ mol^{-1}

The Delocalised Ring of Electrons Stabilises Benzene

The delocalised electron density between carbon atoms in benzene is lower than the bonding electron density between C=C bonds in alkenes. Electrophiles are less attracted to benzene than to alkenes.

Benzene does not easily undergo addition reactions because this would result in the loss of the stabilising effect of the delocalised ring of electrons.

Typical reactions of benzene are **electrophilic substitution** reactions where an electrophile swaps with an H on the ring. This means that the delocalised ring of electrons is retained in the product.

A strong electrophile is needed to attack the stable benzene ring. Electrophilic substitution of benzene uses a catalyst.

Benzene ⟶ Nitrobenzene

benzene + HNO_3 $\xrightarrow{H_2SO_4}$ nitrobenzene (NO_2) + H_2O

The electrophile is NO_2^+, which is formed by the reaction of concentrated nitric and sulfuric acid (known as a nitrating mixture).

$$HNO_3 + 2H_2SO_4 \longrightarrow NO_2^+ + 2HSO_4^- + H_3O^+$$

The NO_2^+ accepts two electrons from the delocalised ring of benzene to form an unstable intermediate.

benzene + NO_2^+ ⟶ intermediate (+, H, NO_2)

The delocalised ring of electrons regenerates using the electrons from the C—H bond.

intermediate ⟶ nitrobenzene (NO_2) + H^+

The catalyst is regenerated: $H^+ + HSO_4^- \longrightarrow H_2SO_4$

If the temperature is kept below 55°C, only one NO_2 substitutes onto the ring. Above 55°C multiple substitutions occur.

Nitration of benzene is an important step in making more complex organic molecules. The nitro group can be reduced to an amine using Sn and hydrochloric acid. Phenylamines are the starting point for the manufacture of many compounds including explosives and dyes.

Benzene ⟶ Halobenzene

The electrophile is a positively charged halogen ion generated with a catalyst.

$$\text{e.g. } Br_2 + FeBr_3 \longrightarrow Br^+ + FeBr_4^-$$

Fe alone can be used as a pre-catalyst since it first reacts with the halogen $3Br_2 + 2Fe \longrightarrow FeBr_3$ to form the catalyst.

benzene + Br^+ ⟶ intermediate (+, Br, H) ⟶ bromobenzene (Br) + H^+

This type of catalyst is known as a halogen carrier.

The lone pair electrons in the halogen are able to overlap with the p orbital electrons in the benzene ring. This gives the C—X, where X is the halogen, in an aromatic compound a higher bond enthalpy than a typical C—X in a non-aromatic compound.

Benzene ⟶ Alkylbenzene

A haloalkane is converted to a positive alkyl group by a halogen carrier.

$$R{-}Cl + AlCl_3 \longrightarrow R^+ + AlCl_4^-$$

benzene + CH_3Br $\xrightarrow{AlCl_3}$ methylbenzene (CH_3) + HBr

Benzene ⟶ Aromatic Ketone

Acyl chlorides are used to add an acyl group to benzene giving a ketone.

$$RCOCl + AlCl_3 \longrightarrow RCO^+ + AlCl_4^-$$

$$C_6H_6 + R{-}COCl \xrightarrow{AlCl_3} C_6H_5COR + HCl$$

Adding a Second Substituent to a Benzene Ring

The presence of one group on a benzene ring determines which H will be substituted for a second group. Some substituents have a tendency to withdraw electrons from the benzene ring, e.g. the NO_2 group.

When a second electrophile attacks the ring it is directed into the 3 position, and the third is directed into the 5 position.

$$C_6H_5NO_2 \longrightarrow 1,3,5\text{-}C_6H_3(NO_2)_3$$

Some substituents are electron donating to the benzene ring, e.g. —OH and —NH_2 groups. These direct a second electrophile into the 2 and 4 positions.

$$C_6H_5OH \longrightarrow 2,4,6\text{-}Cl_3C_6H_2OH$$

Benzene ⟶ Benzenesulphonic Acid

Refluxing benzene with concentrated sulfuric acid produces SO_3. This is the electrophile that begins the substitution.

$$C_6H_6 + H_2SO_4 \longrightarrow C_6H_5SO_3H + H_2O$$

Combustion of Benzene

Benzene burns completely to form carbon dioxide and water but the high C : H ratio means that a large amount of O_2 is required. As a result it burns with a sooty flame full of carbon particles, indicating incomplete combustion.

Phenol

Phenol is a benzene ring with an —OH group directly attached to the ring.

The lone pair electrons on the oxygen atom are in p orbitals at right angles to the plane of the benzene ring.

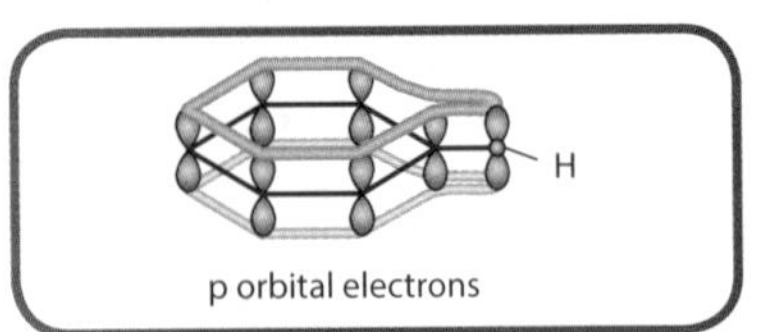

They are able to join with the delocalised electron ring of the benzene, increasing the electron density in the ring.

This makes it easier for electrophiles to attack the ring. All electrophilic substitution reactions happen more readily with phenol than with benzene.

Phenol Decolourises Bromine Water

This is a substitution reaction, unlike the addition reaction of alkenes with bromine.

$$C_6H_5OH + 3Br_2 \longrightarrow 2,4,6\text{-}Br_3C_6H_2OH + 3HBr$$

2,4,6-tribromophenol precipitates as a white solid.

QUICK TEST

1. Draw the structure of 1,4-dichlorobenzene.
2. What type of reactions are typical for benzene?
3. Give the reagents and conditions for the nitration of benzene.
4. Which catalyst should be used to make methylbenzene from benzene?
5. Why does benzene burn with a sooty flame?
6. Why is phenylmethanol not a phenol?

PRACTICE QUESTIONS

1. Benzene is reacted with concentrated sulfuric acid and concentrated nitric acid at 15°C. The major organic product is **[1 mark]**

   A nitrobenzene ☐ **B** 2, 4, 6-trinitrobenzene ☐ **C** phenylamine ☐ **D** 1,3,5-trinitrobenzene ☐

2. Phenols are more subject to electrophilic substitution than benzene. This is because **[1 mark]**

   A the electronegative oxygen withdraws electrons from the benzene ring ☐

   B the lone pair of electrons on oxygen donates electrons to the benzene ring ☐

   C the O—H group of phenol is slightly acidic ☐

   D hydrogen bonding in the —OH disrupts the stability of the delocalised electron ring. ☐

3. Benzene has the formula C_6H_6. Cyclohexene has the formula C_6H_{10}.

   a) Compare the structures of benzene and cyclohexene. **[4 marks]**

   b) Describe the effect of shaking benzene and cyclohexene with bromine water. **[2 marks]**

   c) Give a reason for the difference in response. **[3 marks]**

   d) Name the type of reaction when cyclohexene reacts with bromine water. **[1 mark]**

   e) Name the type of reactions that benzene most readily undergoes. **[1 mark]**

4. The table below shows the structures of some aromatic compounds.

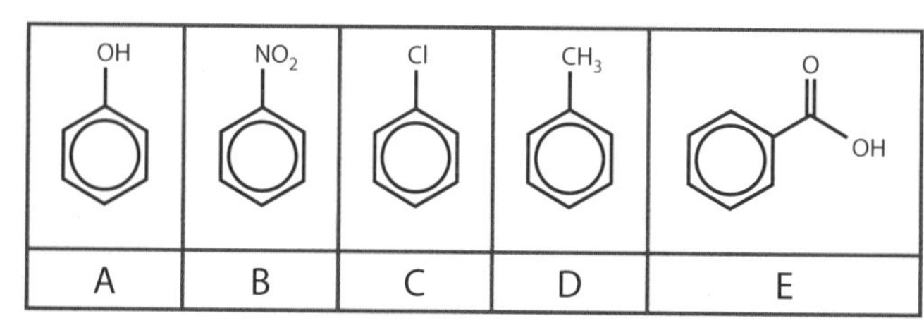

   Select the molecule from the table, A–E, to match each description.

   a) A molecule that could be made from benzene and a chloroalkane. **[1 mark]**

   b) A molecule that is likely to undergo substitution onto the ring in the 3 and 5 positions. **[1 mark]**

   c) A molecule that is formed from benzene using concentrated sulfuric acid as a catalyst. **[1 mark]**

   d) A molecule that could be formed from the oxidation of methylbenzene. **[1 mark]**

   e) A molecule that reacts with alkalis but not carbonates. **[1 mark]**

5. In the 1860s the chemist Kekulé proposed a structure for benzene. He claimed the inspiration came after dreaming about a snake biting its own tail.

   Describe the evidence that suggests that this structure for benzene is not correct. **[4 marks]**

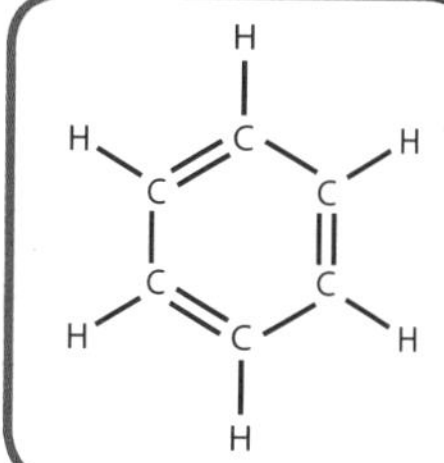

Organic Nitrogen Compounds

Amines

Amines contain the functional group:

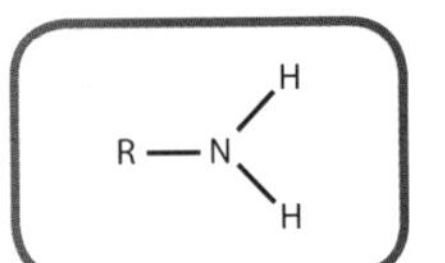

They are named by naming the whole molecule and adding the suffix -amine.

$$H-CH_2-CH_2-CH(CH_3)-CH_2-NH_2$$

2-methylbutylamine

$$R-\overset{\delta-}{N}(H)-\overset{\delta+}{H}\cdots\overset{\delta-}{:O}H_2 \qquad R-\overset{\delta-}{N}\cdots\overset{\delta+}{H}-\overset{\delta-}{O}-H$$

Amines are classified as primary, secondary or tertiary according to the number of hydrogen atoms on the nitrogen of the amine.

Primary amine (1°) 1 alkyl group and 2 hydrogens	$R-NH_2$
Secondary amine (2°) 2 alkyl groups and 1 hydrogen	$R-NHR'$
Tertiary amine (3°) 3 alkyl groups	$R-NR'R''$
Quaternary ammonium ion 4 alkyl groups and a positive charge	$[R-NR'R''R''']^+$

1° and 2° Amines Can Hydrogen Bond

1° and 2° amines can hydrogen bond to each other and have a higher boiling point than alkanes with comparative mass.

1° and 2° amines are able to form hydrogen bonds with water. Ammonia, with no alkyl groups, is very soluble; amines become less soluble as the size of the alkyl group increases.

Amines are Bases

The nitrogen atom of an amine has a lone pair of electrons. This can accept a proton to form a substituted ammonium ion.

$$R-NH_2: + H_2O \rightleftharpoons R-NH_3^+ + OH^-$$

$$R-NH_2: + H^+ \rightleftharpoons R-NH_3^+$$

When a primary amine forms an ammonium ion the nitrogen gains a positive charge. The charge can be spread across the alkyl group and this stabilises the ion, making a stronger base.

When ammonia forms an ammonium ion there is no alkyl group to share the charge so ammonia is a weaker base than a primary amine.

In aromatic amines, the lone pair of electrons on the nitrogen becomes part of the delocalised ring of electrons on the benzene ring. This makes them less available to accept a proton. Aromatic amines are weaker bases than ammonia.

$$C_6H_5NH_2 \quad H^+ \rightleftharpoons C_6H_5NH_3^+$$

Amines React with Acids to Form Salts

Like all bases, amines react with acids to form salts.

$$CH_3CH_2NH_2 + HCl \longrightarrow CH_3CH_2NH_3Cl$$

Amines are Ligands

The lone pair of electrons on the N of an amine can be donated towards a central metal ion, e.g.

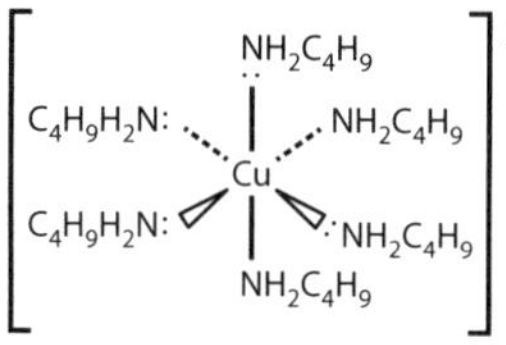

$$[Cu(NH_2C_4H_9)_6]^{2+}$$

$NH_2C_4H_9$ $C_4H_9H_2N$: Cu $NH_2C_4H_9$ $C_4H_9H_2N$: $NH_2C_4H_9$ $NH_2C_4H_9$ 2+

Preparation of Amines

Haloalkane ⟶ Amines

This is a nucleophilic substitution reaction.

$$R{-}Br + 2NH_3 \longrightarrow RNH_2 + NH_4Br$$

The amine acts as a nucleophile by donating a pair of electrons to the δ^+ carbon of the haloalkane.

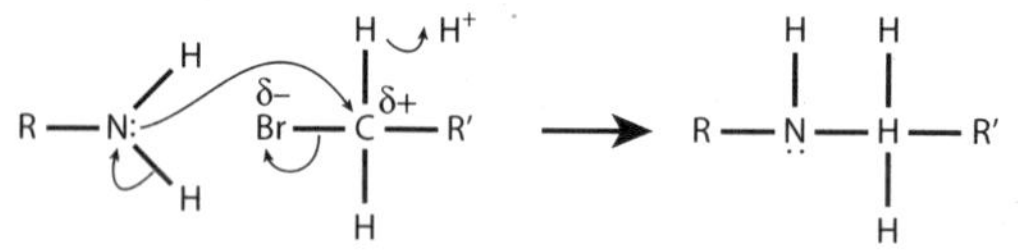

R—N(H)(H): δ– Br—C(H)(H)—R′ δ+ H⁺ ⟶ R—N(H)—H(H)(H)—R′

The reaction takes place in a sealed tube to prevent the ammonia gas from escaping. To prepare a primary amine, excess ammonia is used, otherwise the substitution on the nitrogen continues. The amine produced can still act as a nucleophile since it still has a lone pair of electrons.

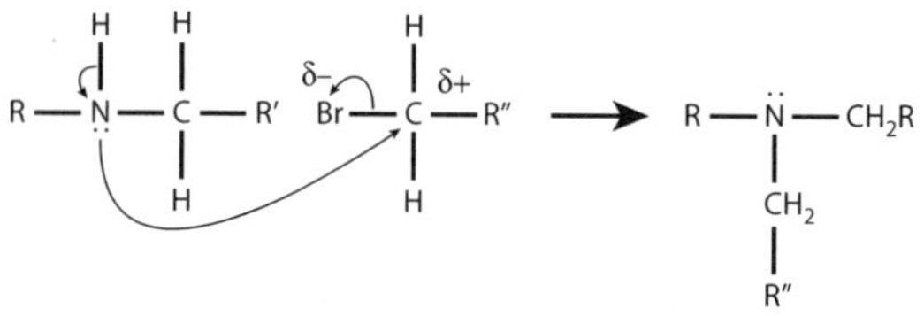

R—N(H)—C(H)(H)—R′ δ– Br—C(H)(H)—R″ δ+ ⟶ R—N(CH₂R″)—CH₂R

The reaction can continue to form a quaternary ammonium ion if sufficient haloalkane is available.

$$R{-}NR_2 + R\text{-}Br \longrightarrow [NR_4]^+ Br^-$$

Nitrile ⟶ Amine

Nitriles are reduced to amines by reaction with $LiAlH_4$ in ether:

$$RCN + 4[H] \longrightarrow RCH_2NH_2$$

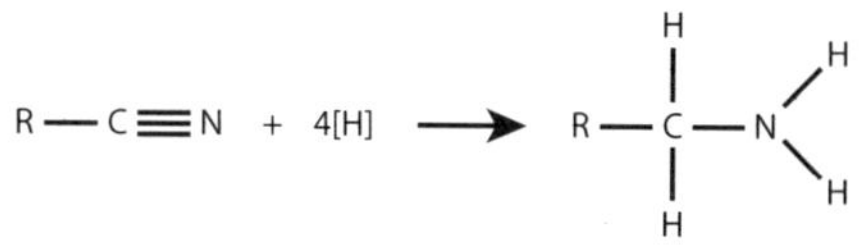

R—C≡N + 4[H] ⟶ R—C(H)(H)—N(H)(H)

Preparation of Phenylamine

Aromatic amines can be prepared by reduction of nitrobenzenes with Sn and concentrated HCl(aq). ([H] represents the reducing agent.)

$$C_6H_5NO_2 + 4[H] \longrightarrow C_6H_5NH_2 + 2H_2O$$

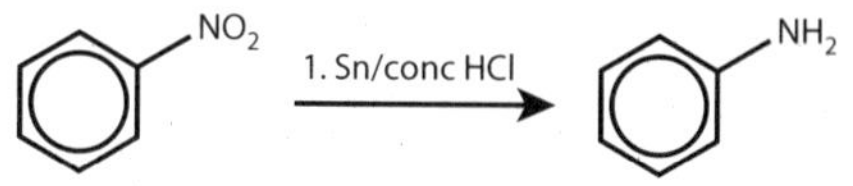

NO_2 1. Sn/conc HCl ⟶ NH_2

Amides

Amides contain the functional group:

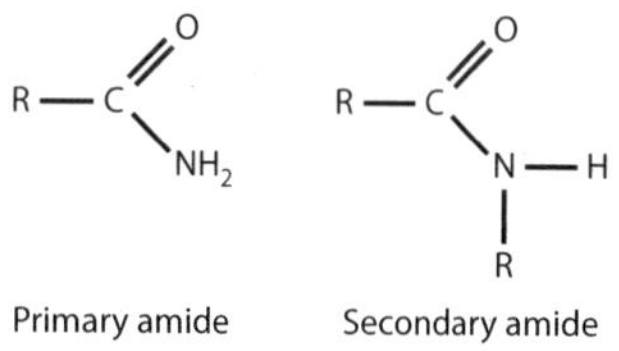

R—C(=O)—NH_2 R—C(=O)—N(R)—H

Primary amide Secondary amide

Carboxylic acids and acyl chlorides react with primary amines to give secondary amides. The reaction begins with addition to the C=O and finishes with elimination; known as an **addition–elimination** reaction.

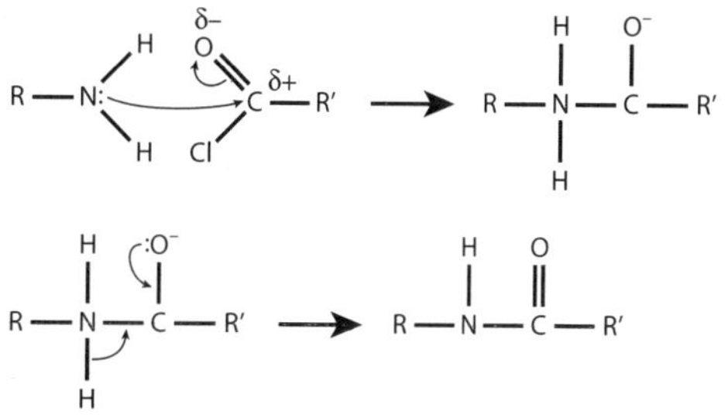

R—N(H)(H): δ– O δ+ C(Cl)—R′ ⟶ R—N(H)(H)—C(O⁻)—R′

R—N(H)(H)—C(:O⁻)—R′ ⟶ R—N(H)—C(=O)—R′

Amides can be hydrolysed with acid or alkali.

$$R{-}C({=}O){-}N(H){-}R' \xrightarrow{H_2O/H^+} R{-}C({=}O){-}OH + H{-}\overset{+}{N}H_2{-}R'$$

$$R{-}C({=}O){-}N(H){-}R' \xrightarrow{H_2O/OH^-} R{-}C({=}O){-}O^- + H_2N{-}R'$$

Acid hydrolysis produces an ammonium ion and a carboxylic acid. Alkaline hydrolysis produces an amine and a carboxylate ion.

Amines Are Used to Prepare Azo Compounds

Aromatic amines such as phenylamine react with nitric(III) acid (HNO_2) to form diazonium ions.

The nitric(III) acid is prepared in situ by mixing $NaNO_2$ with hydrochloric acid. The temperature must be kept below 5°C to prevent the diazonium ion from breaking down to N_2.

$$NaNO_2 + HCl \longrightarrow HNO_2 + NaCl$$

$$C_6H_5NH_2 + HNO_2 + HCl \longrightarrow C_6H_5\overset{+}{N}{\equiv}N + 2H_2O + Cl^-$$

The diazonium ion can then be reacted with a phenol or aromatic amine to form an azo compound under alkaline conditions. The diazonium ion is an electrophile, which reacts with the benzene ring in a substitution reaction.

$$C_6H_5\overset{+}{N}{\equiv}N + C_6H_5OH \longrightarrow C_6H_5{-}N{=}N{-}C_6H_4{-}OH$$

The reaction is known as a coupling reaction and the aromatic compound that is added to the diazonium ion is called a **coupling agent**.

Azo compounds are highly coloured and are used as dyes.

The Azo Group Is Coloured Because of Delocalisation

The azo group —N=N— in an azo compound is stabilised by the two benzene rings on either side. The delocalised ring of the benzene, the lone pairs on the two nitrogen atoms of the azo group and the π bond electrons between the two nitrogen atoms all join together to form an extended delocalised system.

In the delocalised system the difference between electron energy levels is small. This means that the energy provided by photons in the visible spectrum can be absorbed by electrons that are then excited to the higher energy level.

These photons are removed from white light so the remaining photons will be the colour observed. This means the complementary colour is seen. Different substituents on the benzene rings cause slight differences in the energy levels and so give different colours.

$$HO{-}C_6H_4{-}N{=}N{-}C_6H_5$$

Orange

$$H_2N{-}C_6H_4{-}N{=}N{-}C_6H_5$$

Yellow

The —N=N— bond is responsible for that fact that the molecule is coloured and is known as the **chromophore**.

QUICK TEST

1. Why are amines water soluble?
2. Draw the general structure of a secondary amine.
3. Why can amines act as ligands?
4. Give the reactants for the reduction of nitrobenzene to phenylamine.
5. What type of reaction is the conversion of a haloalkane to an amine?
6. What is a chromophore?

PRACTICE QUESTIONS

1. Three nitrogen containing compounds are ammonia NH_3, propylamine $C_3H_7NH_2$ and phenylamine $C_6H_5NH_2$.

   Placed in order of increasing strength as bases the order would be **[1 mark]**

A	NH_3	<	$C_3H_7NH_2$	<	$C_6H_5NH_2$	☐
B	NH_3	<	$C_6H_5NH_2$	<	$C_3H_7NH_2$	☐
C	$C_6H_5NH_2$	<	NH_3	<	$C_3H_7NH_2$	☐
D	$C_3H_7NH_2$	<	NH_3	<	$C_6H_5NH_2$	☐

2. Which of the following is not a way of preparing amines? **[1 mark]**

   A Reducing a nitrile with $LiAlH_4$. ☐

   B Heating a bromoalkane with ammonia. ☐

   C Hydrolysing a secondary amide. ☐

   D Reacting an acyl chloride with ammonia. ☐

3. Propylamine can be prepared by reacting bromopropane with ammonia in ethanol.

   a) Write an equation for this reaction. **[1 mark]**

   b) Outline the mechanism for this reaction, and name the type of reaction. **[3 marks]**

   c) The reaction is likely to result in a mixture of products. Give a reason for this and suggest a way in which products other than propylamine can be minimised. **[3 marks]**

4. Carboxylic acids react with primary amines to give secondary amides via an addition–elimination reaction.

   a) Draw the structure of the functional group of a primary amine and a primary amide. **[2 marks]**

   b) Draw the mechanism for the formation of an amide from methylamine and ethanoic acid. **[3 marks]**

   c) Suggest why this reaction is unlikely to occur in acidic conditions. **[1 mark]**

5. Azo compounds are coloured substances that are often used as dyes. They are made in a two-step reaction.

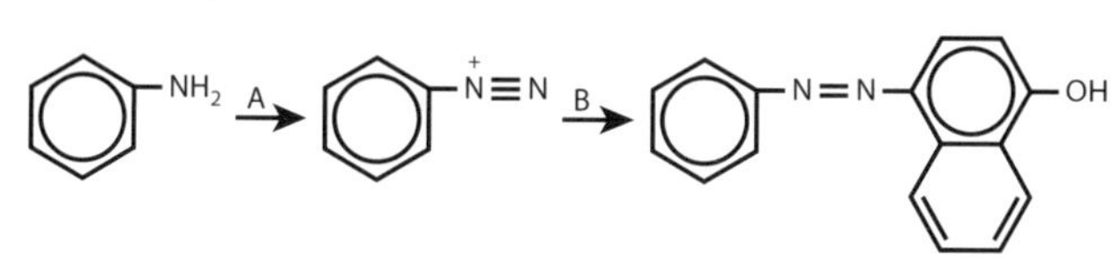

   a) Give the reagents and conditions needed for the first step. **[3 marks]**

   b) What type of reaction completes the reaction in the second step? **[1 mark]**

   c) Draw the structure of the molecule needed for the second step. **[1 mark]**

Amino Acids and Proteins

Amino acids are the monomer units of proteins. They contain both an amine and a carboxylic acid group.

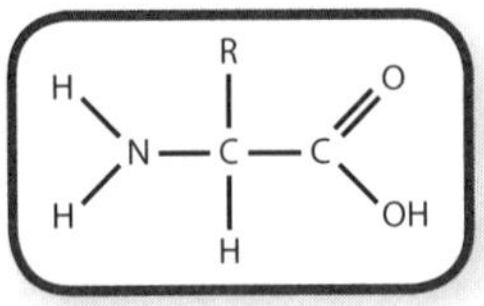

2-amino acids or α amino acids have the amine and carboxylic acid group both attached to the second carbon atom. They have non-systematic names according to their R group.

In naturally occurring amino acids, the R group may be one of 20 different structures. Among the different structures are aromatic groups, carboxylic acid groups, amine groups, amide groups and simple alkyl chains. Two amino acids contain sulfur atoms.

Amino Acids in Different pH Environments

The acid group in amino acids can donate a proton.

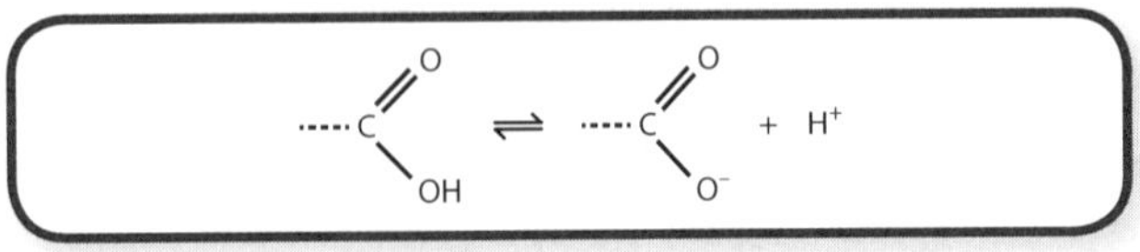

In low pH environments the high [H^+] moves the position of equilibrium to the left. As the pH increases, the position of equilibrium moves to the right and the acid group has a negative charge.

The amine group can accept a proton.

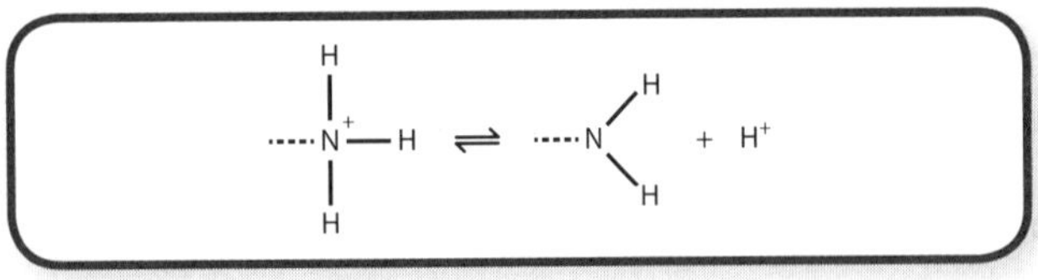

In low pH environments the amine group accepts a proton and gains a positive charge. As the pH rises the position of equilibrium moves to the right and the amine loses the proton.

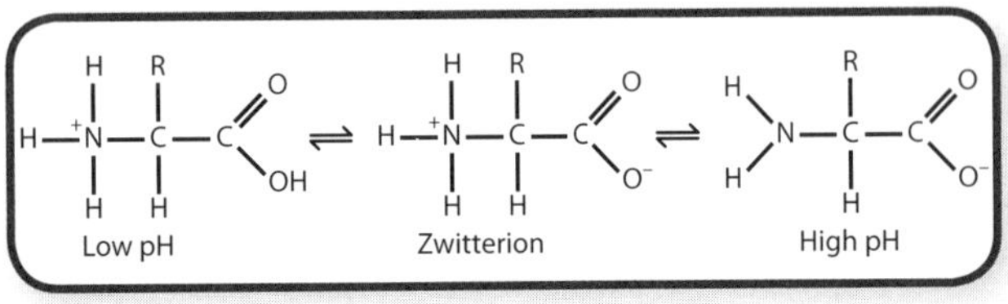

Low pH Zwitterion High pH

At neutral pH both the acid and the amine group are charged. The H^+ donated by the carboxyl group is accepted by the amine group forming an internal ion known as a zwitterion.

Simple molecular structures have low melting points. Amino acids have relatively high melting points, which can be explained by ionic bonding between zwitterions.

The pH at which there is no overall charge on the amino acid is known as the **isoelectric point**. Different amino acids have different isoelectric points due to presence or absence of charged groups in their different R groups.

2-Amino Acids Show Optical Isomerism

A carbon atom with four different groups attached is called a **chiral centre**, chiral carbon or asymmetric carbon. Molecules with a chiral carbon have non-superimposable mirror images.

Note: It is not 4 different *atoms* attached to the chiral carbon but 4 different *groups*, including everything covalently bonded together.

Molecules with a chiral centre can exist in two isomers that are non-superimposable mirror images of each other.

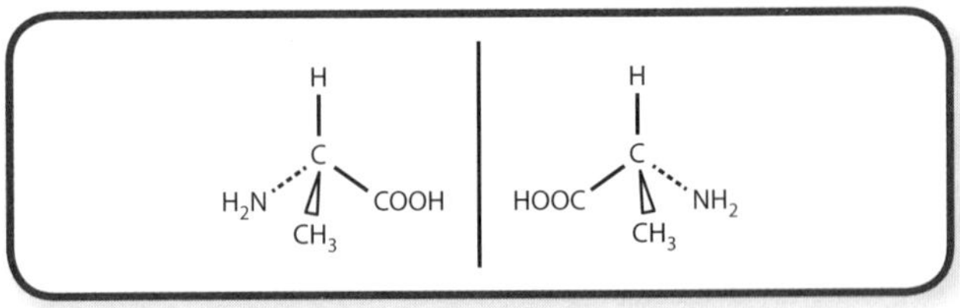

The two isomers are called **enantiomers** or optical isomers. One is the D-isomer and the other the L-isomer.

Optical isomerism is a form of stereoisomerism. Isomers have the same molecular formula and the atoms of optical isomers are in the same order but have a different arrangement in space.

When plane-polarised light is shone through a sample of a single enantiomer it causes the plane of polarisation to rotate. The higher the concentration the more the rotation. The other enantiomer rotates the plane of polarised light in the opposite direction. Solutions that rotate the plane of plane-polarised light are said to be optically active.

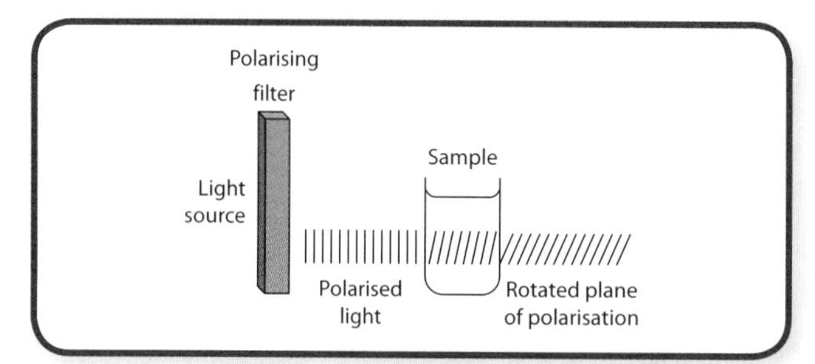

A 50/50 mixture of enantiomers is called a **racemic mixture** and does not rotate the plane of polarised light. As there are the same number of each isomer, the effect of light rotation by one isomer is cancelled out by the other. It is described as optically inactive.

All naturally occurring amino acids except glycine (where R = H) have a chiral centre at the 2-carbon atom. Naturally occurring amino acids made by enzyme action of living organisms are all the L-isomer.

The Peptide Bond

Amino acids can join together to form a secondary amide (see page 81 secondary amides).

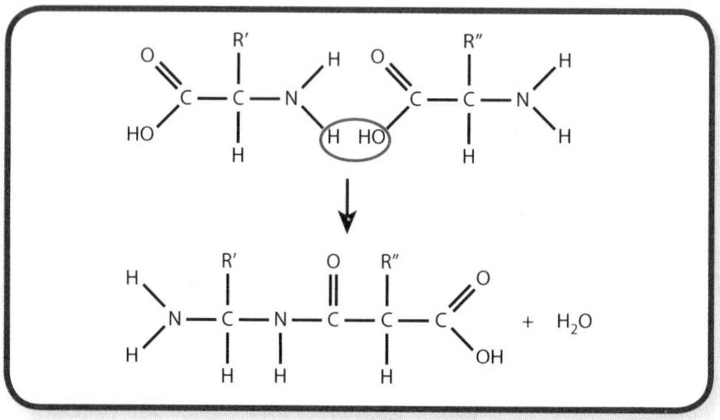

$$+ H_2O$$

This is a condensation reaction. Two molecules join together with the elimination of a smaller molecule.

Two amino acids joined together are known as a **dipeptide**. The amide bond between naturally occurring amino acids is known as a peptide bond.

There are two possible peptides that can be made from two different amino acids. The dipeptide below links the two amino acids in the reverse order.

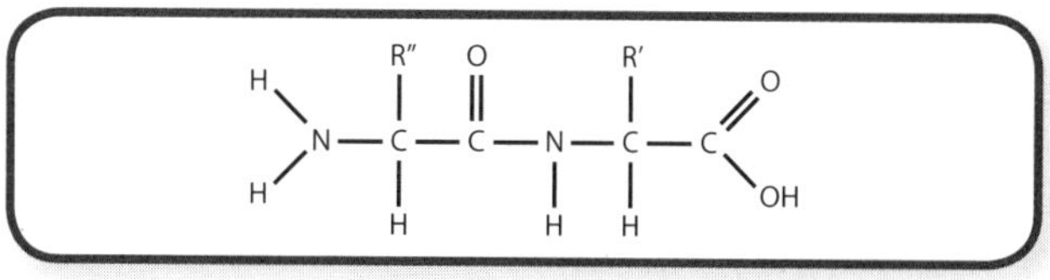

A dipeptide has a free amino group and a free acid group so more amino acids can join on either end. Many amino acids joined together via peptide bonds are called a **polypeptide**.

Proteins

Proteins are long polypeptide chains formed from amino acids linked together in a specific order.

Primary Structure Is the Sequence of Amino Acids

The order in which the 20 different amino acids are linked by amide bonds is known as the primary structure of a protein. It is determined by the genetic code on DNA. Covalent bonds hold the primary structure together.

Secondary Structure Is Hydrogen Bonding in the Polypeptide Backbone

The peptide bond in the backbone of all polypeptides can hydrogen bond. The peptide bonds can align themselves to a hydrogen bond in two ways:

- **α-helix:** The polypeptide chain forms a helix so that every fourth peptide bond has a N—H group in position to bond with a C=O group.

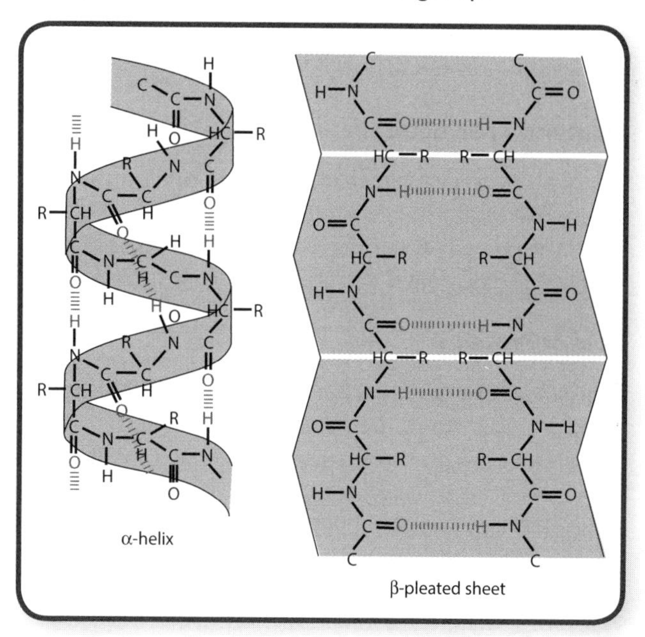

α-helix β-pleated sheet

- **β-pleated sheet:** The polypeptide chain turns back on itself so that the backbone is running parallel to itself and the peptide groups are in the correct position to H-bond.

Hydrogen bonds hold the secondary structure together.

Tertiary Structure

The tertiary structure is the three-dimensional shape of the protein held together by interactions between the R groups.

R groups that contain amine or acid groups can hydrogen bond with each other if unionised or form ionic bonds if ionised, for example:

Lysine $R{=}(CH_2)_4{-}NH_2$ Aspartic acid $R{=}CH_2COOH$

R groups that are polar tend to dissolve into water while non-polar groups cluster together and form instantaneous dipole–induced dipole bonds. This results in the non-polar groups bonding together in the centre of the structure while the polar R groups interact with water on the edges.

The amino acid cysteine has the R group $-CH_2-SH$. When two cysteine R groups from different parts of the polypeptide chain come together a reduction reaction can form a covalent bond.

$$-CH_2-SH \quad HS-CH_2- \longrightarrow -CH_2-S-S-CH_2-$$

These many interactions between R groups hold the structure into a three-dimensional shape that is required for the function of the protein.

Quaternary Structure

Many proteins contain more than one polypeptide chain. The quaternary structure is the way that different polypeptide chains join to form the functioning protein.

Role of Proteins

Proteins are an essential part of all living organisms. They provide structure in hair, nails, bone and muscles. Some hormones are proteins, e.g. insulin and growth hormone. Enzymes are proteins. Enzymes are biological catalysts. For kinetics of enzyme-controlled reactions see page 88.

Hydrolysis of Proteins

The peptide bonds in proteins can be hydrolysed in moderately concentrated acid or alkali. If alkali is used the salt of the amino acid is produced and if acid is used the ammonium ion is produced. Neutralising the acid or alkali gives the zwitterion form of the amino acids.

The component amino acids can be separated and identified using paper or thin layer chromatography (see page 101).

QUICK TEST

1. Draw the structure of the amino acid that has CH_3 as its R group.
2. What is a zwitterion?
3. Why is the formation of a peptide bond called a condensation reaction?
4. What is an enantiomer?
5. What is meant by optically active?
6. What bonding holds the tertiary structure of proteins in place?

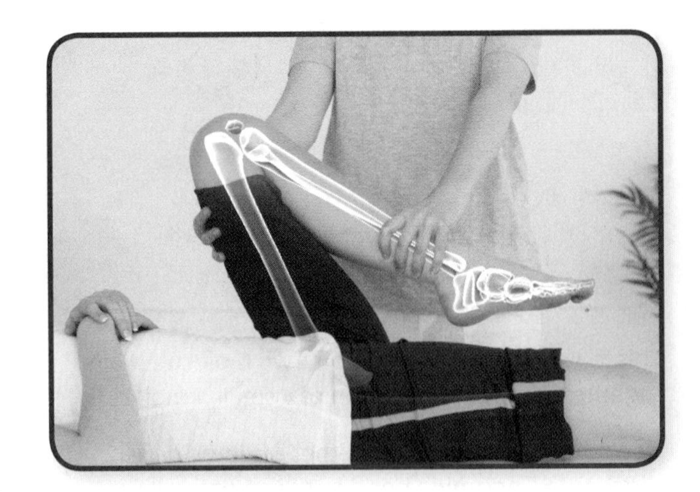

PRACTICE QUESTIONS

1. Which of the following is **not** true of amino acids? **[1 mark]**

   A They are simple covalent structures as solids. ☐

   B They are the monomer units of polypeptides. ☐

   C They can exist as both negative and positive ions. ☐

   D They are water soluble. ☐

2. Which is true of molecules that are optical isomers? **[1 mark]**

   A They have the same formula but a different order of atoms. ☐

   B They have the same formula but different functional group. ☐

   C They have four different atoms attached to one carbon. ☐

   D They are optically active. ☐

3. Alanine is a 2-amino acid.

   H_3C O H_3C OH NH_2

   Alanine

   a) Copy the diagram and identify the chiral centre in alanine. **[1 mark]**

   b) Draw a diagram of the two optical isomers of alanine. **[2 marks]**

   c) Alanine exists as a zwitterion. Draw a diagram of the zwitterion structure of alanine. You may use R to represent the side group. **[2 marks]**

   d) Naturally occurring alanine exists as the L-isomer. How could you demonstrate that a sample of alanine contained the L-isomer rather than the D-isomer? **[2 marks]**

   e) What is meant by a racemic mixture? **[1 mark]**

4. Proteins are polymers of amino acids that carry out many important functions in living organisms. The structure of a protein is vital to its function and is described on different levels from primary to quaternary structure. The general structure of amino acids is shown below.

   R H O N—C—C H OH H

   a) Show the way that two amino acids join together to form a dipeptide. **[2 marks]**

   b) What is meant by the primary structure of a protein? **[1 mark]**

   c) What type of bonding holds the secondary structure of proteins in place? **[1 mark]**

   d) Explain why the primary structure of a protein has an important influence on the eventual overall shape of the protein. **[3 marks]**

Enzymes and DNA

Enzymes are proteins that function as catalysts for biological reactions. Each enzyme catalyses a specific reaction.

Enzymes have a specific 3-D shape that includes a region known as the **active site**. The shape of the active site is complementary to the shape of the reactants for the reaction to be catalysed. The reactants (known as the **substrate**) fit into the active site and are held in place by intermolecular forces. The reaction is catalysed and the products are released from the active site.

An enzyme catalysed reaction is a multistep reaction.

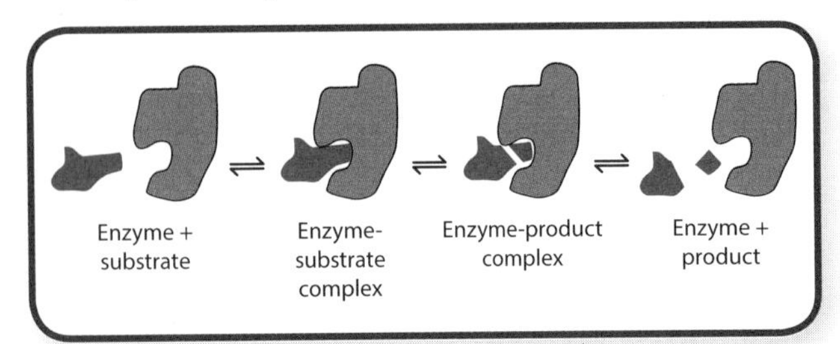

At low substrate concentrations, the first step is the rate-limiting step and the reaction is first order with respect to substrate.

At high substrate concentrations every enzyme active site is occupied and the second step becomes the rate-limiting step. The reaction is zero order with respect to substrate.

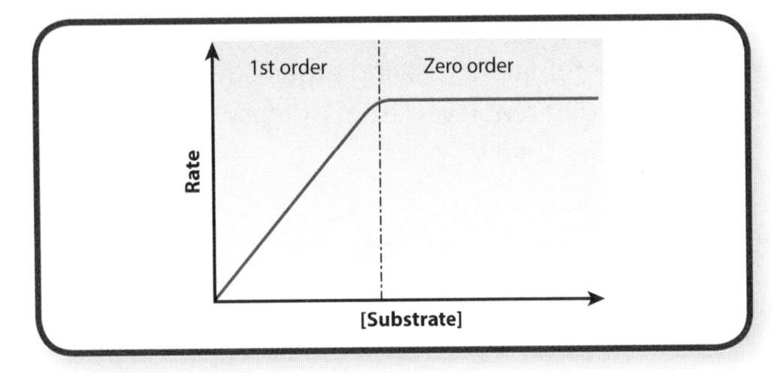

A High Temperature Denatures Enzymes

The shape of the active site is part of the tertiary structure of the protein (see page 86). The tertiary structure is mostly held in place by intermolecular bonding, which is easily broken by heating. For a typical enzyme, heating to temperatures above 40°C results in loss of shape of the active site and permanent loss of activity (denaturation) of the enzyme.

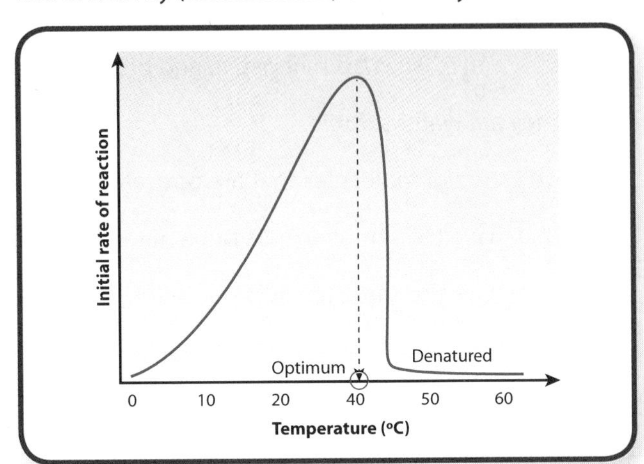

Enzymes Are pH Specific

The substrate is held in the active site via ionic bonding or hydrogen bonding. Changing the pH changes the degree of ionisation of the —NH_2 and —COOH groups in the R groups in the active site and interferes with the substrate binding. This means that enzymes are only effective catalysts within a narrow range of pH values.

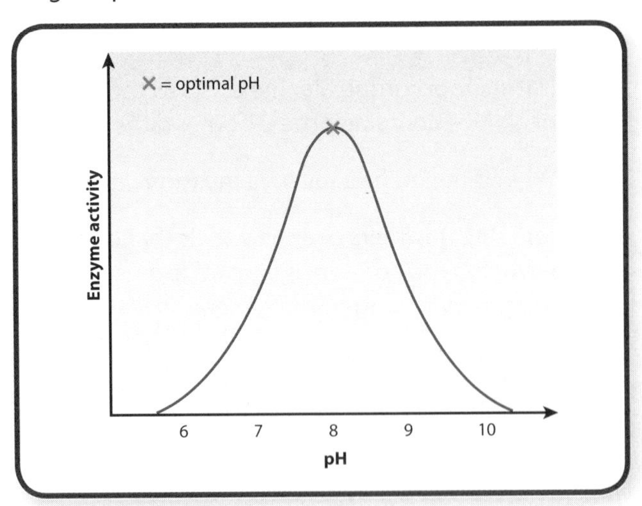

Competitive Inhibitors Fit in the Active Site

Competitive inhibitors are molecules with a similar shape to the substrate that fit into the active site of the enzyme. This prevents substrate from entering the active site and reduces the concentration of useful enzyme.

Some Medicines Work by Binding to Active Sites

Many medicines are enzyme inhibitors. Ibuprofen is a competitive inhibitor of the enzyme cyclooxygenase (COX). This temporarily reduces the synthesis of compounds that result in inflammation and pain.

Molecular recognition is intermolecular bonding between biological molecules that have complementary shapes. A protein with a specific 3-D shape forms a receptor site that can bind to a complementary shaped molecule. This results in a specific action, such as allowing the passage of ions across membranes or triggering a chain of reactions. Examples include enzyme–substrate, neuroreceptor–neurotransmitter, hormone–cell membrane.

Many medicines make use of molecular recognition and bind to specific receptor sites to either increase or inhibit the normal response.

Receptor Sites Are Stereospecific

If a medicine has a chiral centre, only one enantiomer will be complementary to the receptor site. The manufacture of medicines often results in a mixture of enantiomers. At best this reduces the effectiveness of the medicine. At worst, other enantiomers may have damaging side effects.

Enzyme catalysed reactions only produce one enantiomer because the active site of the enzyme is chiral.

The Pharmacophore Conveys the Drug Activity

The part of the molecule that results in the pharmacological action is called the **pharmacophore**. This is the part that enters the receptor site. Chemists can modify parts of the molecule bonded to the pharmacophore to improve the drug. This may change the solubility, increase the efficacy or reduce side effects.

The 3-D shape of some proteins can now be visualised via computer graphics. Computer modelling enables chemists to design medicines that are the correct shape for specific receptor sites.

DNA is the Genetic Material in the Nucleus of Cells

DNA is a condensation polymer whose monomer units are nucleotides.

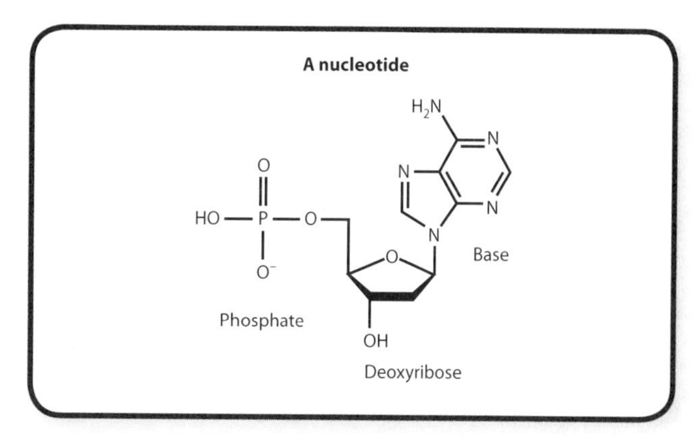

The base can be one of four molecules: adenine, thymine, cytosine or guanine.

A condensation reaction links the —OH group of deoxyribose and the —H of phosphate. This results in a polymer chain of sugar-phosphate with the bases as side groups.

Each base is a planar molecule that can hydrogen bond with only one other.

Adenine Thymine

Guanine Cytosine

DNA Replication Depends on Hydrogen Bonding Between Pairs of Bases

When DNA replicates, the bases from individual nucleotides line up and hydrogen bond with the bases on a DNA strand.

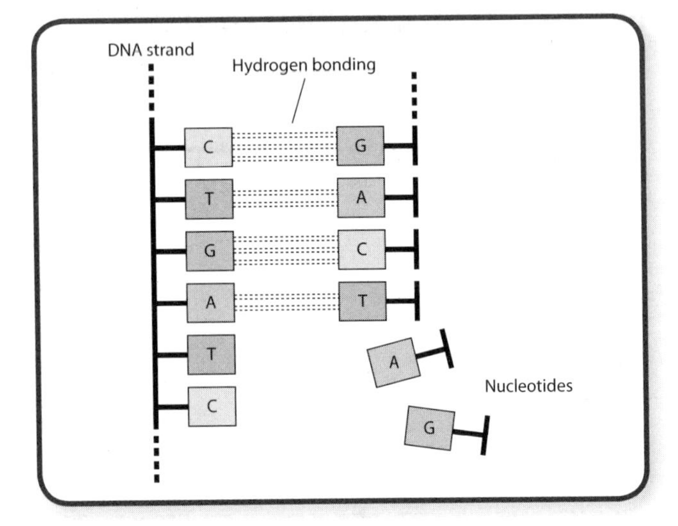

The nucleotides are then linked together by an enzyme catalysed condensation reaction. This forms a new strand of DNA with a complementary sequence of bases to the original strand.

DNA is stored in the nucleus of cells as two strands held together by hydrogen bonds, forming a double helix structure. When a new cell is made the strands separate and each makes a complementary strand.

Cancer Drugs Inhibit DNA Replication

Cancer tissue makes new cells at a rapid rate. Many cancer drugs work by preventing cell reproduction. Cisplatin is a drug that inhibits DNA replication by binding to DNA via coordinate bonds.

A ligand substitution reaction replaces the two chloride ligands with two guanine ligands from the DNA strand. This interferes with the replication of the DNA and so reduces cell replication.

Cancer drugs such as cisplatin are not specific to cancer cells and so also interfere with the replication of other cells. This particularly affects fast replicating cells, such as those lining the gastrointestinal tract, white blood cells and hair follicle cells. An increased chance of inaccurate DNA replication also increases the risk of further cancer.

The Sequences of Bases in DNA Codes for a Protein

Each group of three bases on DNA is a code for an amino acid. A gene is a section of DNA that gives the codes for the primary structure of a protein.

An mRNA copy of the gene is made by base pairing of RNA nucleotides with the DNA strand. RNA is a nucleotide polymer that differs from DNA in using ribose rather than deoxyribose and uracil (a de-methylated form of thymine) in place of thymine.

The code on the RNA ensures the correct primary structure of the protein. A molecule of tRNA carries one amino acid and has three bases available for H-bonding. The three rRNA bases hydrogen bond to three complementary bases on mRNA and the amino acid is added to the protein chain in the correct order.

Hydrogen bonding between the nucleotide bases is the key to all these processes. It ensures that the mRNA is linked together in the correct order and that the tRNA brings the correct amino acid to join the chain.

QUICK TEST

1. List the steps in an enzyme controlled reaction.
2. Why are enzymes pH specific?
3. What is a pharmacophore?
4. Name three parts common to all nucleotides.
5. What type of bonding makes cisplatin effective?
6. How many hydrogen bonds link adenine and thymine?

PRACTICE QUESTIONS

1. The kinetics of an enzyme controlled reaction vary according to the substrate concentration. Which is true? **[1 mark]**

	At high substrate concentration	At low substrate concentration	
A	the reaction is first order with respect to substrate	the reaction is zero order with respect to substrate	☐
B	the reaction is first order with respect to substrate	the reaction is first order with respect to substrate	☐
C	the reaction is zero order with respect to substrate	the reaction is first order with respect to substrate	☐
D	the reaction is zero order with respect to substrate	the reaction is zero order with respect to substrate	☐

2. Which of these statements is **not** true of DNA? **[1 mark]**

   A It forms a double helix held together by hydrogen bonding. ☐

   B It contains the bases adenine, guanine, cytosine and uracil. ☐

   C The nucleotides are joined by a condensation reaction. ☐

   D It carries a code for protein primary structure in the sequence of bases. ☐

3. Enzymes are vital for life but have delicate structures. They are pH specific and can be completely denatured at high temperatures.

   a) Explain why enzymes can only function within a narrow pH range. **[3 marks]**

   b) Use ideas about structure and bonding to explain why enzymes are denatured at high temperatures. **[3 marks]**

   c) Why do some molecules act as enzyme inhibitors? **[3 marks]**

   The reaction between an enzyme and its substrate is an example of molecular recognition.

   d) Give another example of molecular recognition and explain how some medicines are able to take advantage of this process. **[2 marks]**

4. The double helix of DNA is held together by hydrogen bonding between the bases.

   a) Draw the hydrogen bonds that form between guanine and cytosine shown here. **[2 marks]**

   b) How does hydrogen bonding between bases allow DNA to replicate accurately? **[2 marks]**

   c) How does cisplatin interfere with replication? **[2 marks]**

   d) DNA contains a triplet code of bases. What does one triplet code for? **[1 mark]**

Cytosine

Guanine

Polymers

Polymers are very long molecules formed by the linking together of many small molecules known as monomers.

Addition polymerisation has only one product: The monomers of addition polymers are typically alkenes but cyclic compounds without C=C can add together (see below).

Addition polymers made from alkenes are typically very chemically stable and do not degrade readily because of the strength of the bonds in the carbon chain.

Condensation polymerisation has two products: A polymer is produced and also one smaller molecule for each monomer unit added.

Condensation polymers are typically chemically reactive at the points of the monomer joins.

Polyesters

Polyesters are usually formed from diols and dicarboxylic acids.

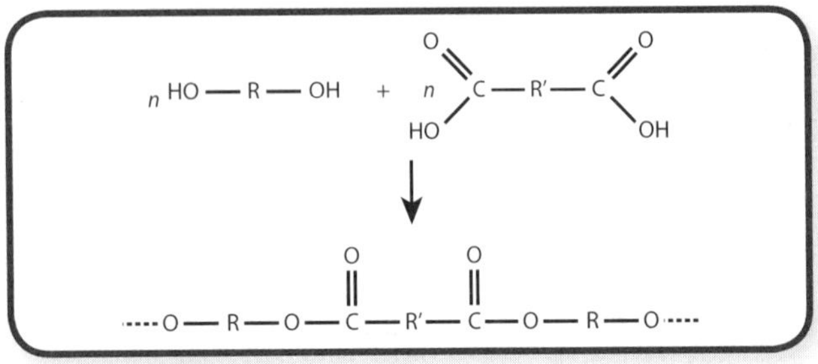

The monomer units in a polyester are held together by an ester group.

The ester group contains a polarised C=O so permanent dipole–permanent dipole bonding can occur between the chains.

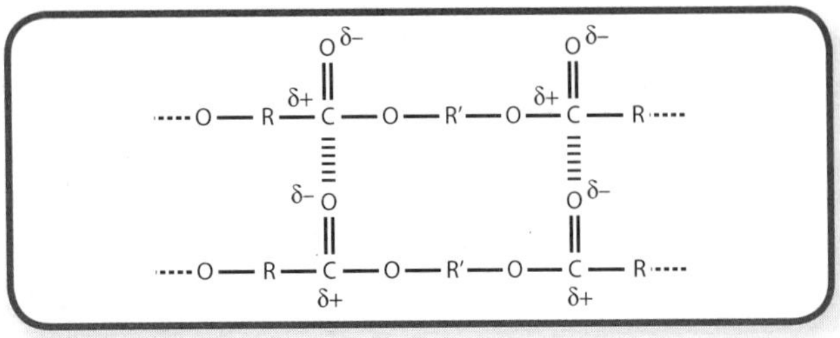

This means that the chains do not slide past each other as easily as polymer chains held together by only instantaneous dipole–induced dipole bonding. As a result, polyesters are stronger.

The ester group is hydrolysed by alkali or acid. Hydrolysis with alkali results in the salt of the dicarboxylic acid and a diol. Hydrolysis with acid results in a dicarboxylic acid and a diol (see page 74 for hydrolysis of esters).

PET, poly(ethyleneterephthalate) is used for plastic bottles and for fibres. It is made from benzene-1, 4-dicarboxylic acid and ethan-1,2-diol.

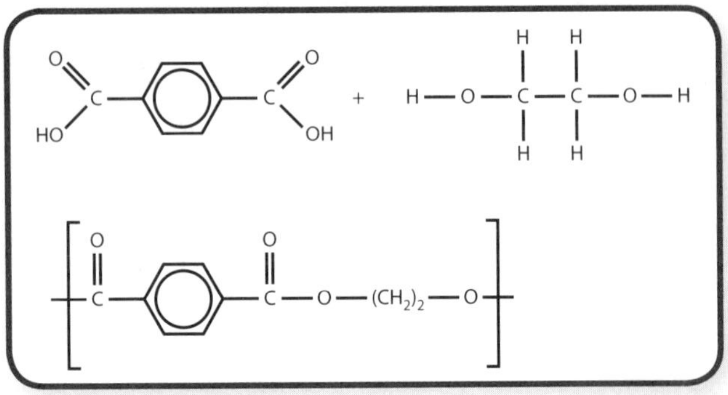

If the dicarboxylic acid monomers are converted to diacyl chlorides the polymerisation is faster. When an acyl chloride and an alcohol form an ester group the molecule eliminated is HCl. This gives a much lower atom economy for the reaction as well as producing dangerous acidic waste.

Polyesters can also be made from molecules that contain both an acid and an alcohol group.

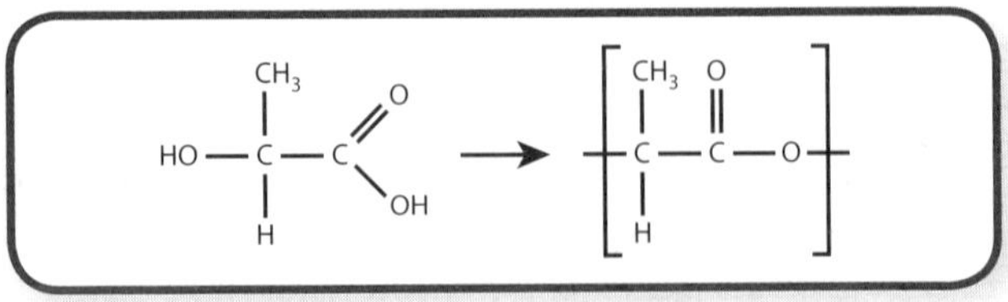

2-hydroxypropanoic acid (lactic acid) forms poly(lactic acid).

Poly(lactic acid) is a biodegradable polyester made from renewable resources but is currently expensive to produce.

Polyamides

Polyamides are formed from a diamine and a dicarboxylic acid. The monomer units are held together by an amide link formed in a condensation reaction.

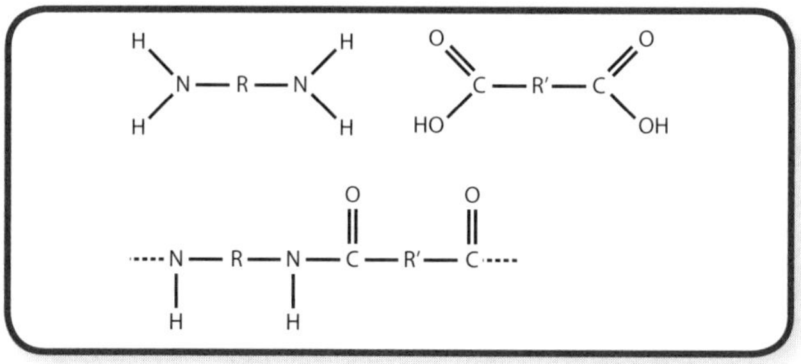

Polyamide chains are able to hydrogen bond with each other.

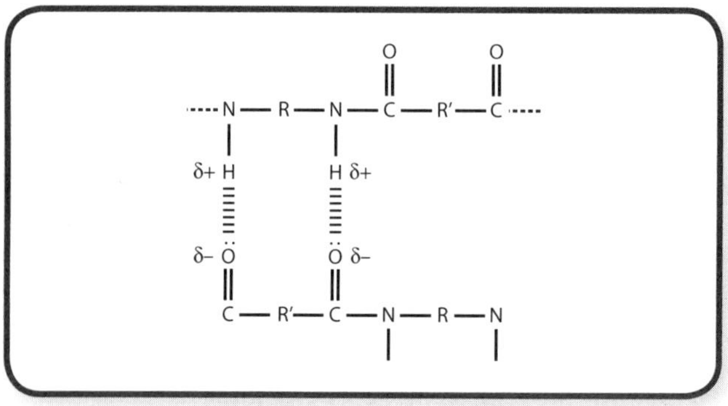

This makes polyamides strong compared to other polymers. If the R group in the acid and in the amine are of the same chain length, the amide links are able to line up and hydrogen bond with each other. If the chain lengths are very different, there is less opportunity for amide groups to hydrogen bond.

Nylons are examples of polyamides. Nylon 6,6 is made from 1,6-diaminohexane and hexanedioic acid.

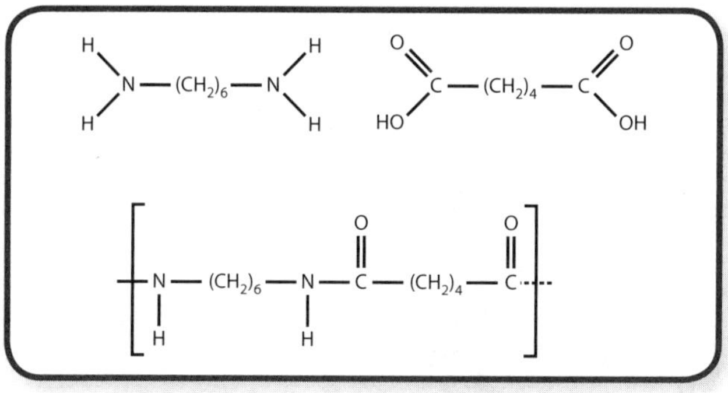

Kevlar is made from two aromatic monomers, benzene-1,4-dicarboxylic acid and benzene-1, 4-diamine.

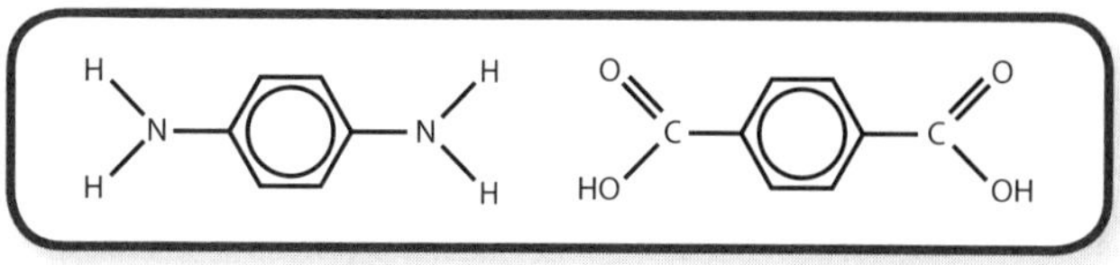

The presence of the benzene ring in the monomers forces the amide groups to align in the same way and this results in very extensive hydrogen bonding forming sheets from the polymer chains.

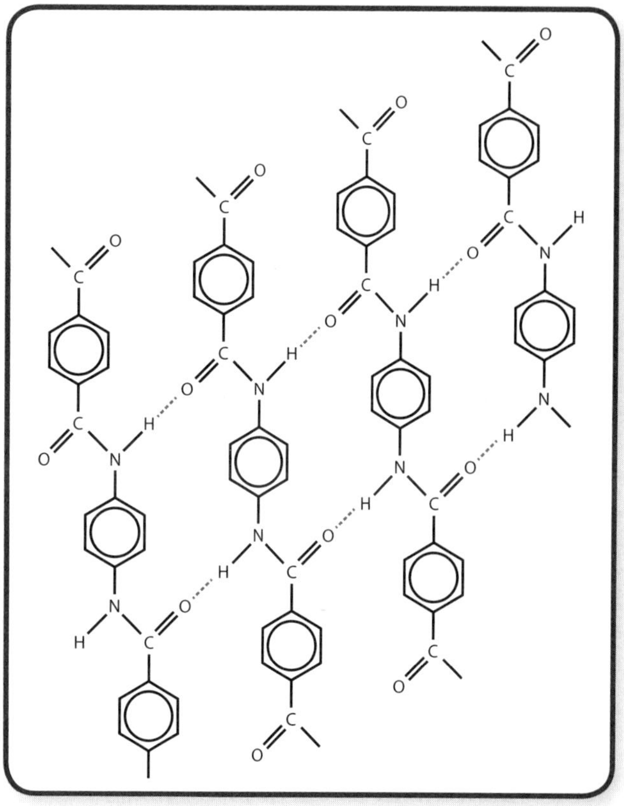

Kevlar is very strong but very expensive to manufacture because concentrated sulfuric acid is used as a solvent to prevent the polymer from precipitating out before it can be aligned into sheets.

Polyamides can be hydrolysed with acid or alkali. Hydrolysis with alkali results in the salt of the dicarboxylic acid and a diamine. Hydrolysis with acid results in the ammonium salt of the diamine and a carboxylic acid.

Polyamides from One Monomer

Nylon 6 is made from one cyclic monomer that already contains an amide bond.

Caprolactam

During manufacture the amide bond breaks, opening the ring and allowing a new amide bond to form between monomers. There is no smaller molecule eliminated so this is not a condensation reaction although it produces a condensation-like polymer.

$$\left[-N(H)-(CH_2)_5C(=O)-\right]$$

Disposing of Polymers

Recycling

Polymers can be collected and recycled. Thermosoftening polymers can be remelted and reshaped but not all polymers will blend together. In order to give a useful product the polymers must be sorted into types. This makes recycling an expensive option. Recycled polymers from general waste are not pure and are of mixed colour. This means the products made are dark in colour and less strong than pure polymers. The most effective recycling of polymers comes from single polymer waste from manufacturers.

Condensation Polymers vs Addition Polymers

The relatively unreactive nature of polyalkenes makes them difficult to break down. There is a strong, non-polar C—C bond between monomer units. Polyalkenes persist for long periods of time if buried in landfill sites. They can be incinerated but there is a risk of producing toxic fumes such as dioxins.

Polyesters and polyamides can be hydrolysed and this makes them easier to dispose of than polyalkenes. They gradually break down in landfill sites.

Polyesters and polyamides can be broken down into their monomer units and the monomers reused to make new polymers. This is currently an expensive option in comparison to making new polymers.

The C═O bond absorbs radiation, which encourages photo-degradation (breakdown in light) of the ester or amide link. Polymers can be treated with UV light to encourage this breakdown.

Polyesters have been shown to break down in soil and a bacterium has been identified that is capable of breaking them down. Thus they are biodegradable. However, the conditions in landfill sites are often not optimal for microbial breakdown of polymers since there are frequently very low oxygen concentrations. The presence of aromatic groups such as in PET makes them less susceptible to biodegradation.

QUICK TEST

1. What is the strongest type of intermolecular bonding between polyester chains?
2. What is the disadvantage of using diacylchlorides as monomers?
3. Why is poly(lactic acid) regarded as a 'green' polymer?
4. Why are polyamides stronger than polyalkenes?
5. Why are polyalkenes difficult to break down?
6. State a danger from incinerating polymers.

PRACTICE QUESTIONS

1. A section of a polymer chain is shown below.

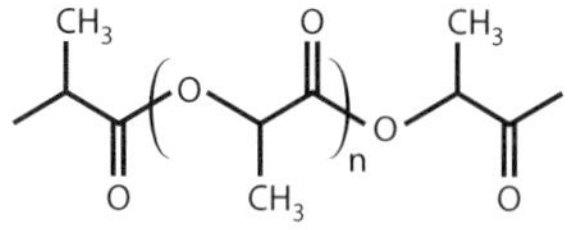

What is true of this polymer? **[1 mark]**

A It is made from a dicarboxylic acid and a diol. ☐

B The strongest intermolecular force between polymer chains is hydrogen bonding. ☐

C It is more easily degraded than polyethene. ☐

D It is an addition polymer. ☐

2. Nylon 6,6 was one of the first nylons to be manufactured. Which statement is **not** true of nylon 6,6? **[1 mark]**

A It is a polyester. ☐

B It can be hydrolysed by moderately concentrated acid. ☐

C The monomer units from which it is made are hexane-1,6-dioic acid and 1,6-diaminohexane. ☐

D It is a condensation polymer. ☐

3. Polyamides are polymers that were originally designed to be a man-made alternative to silk. They are condensation polymers that are mostly made from two different monomers that link together in an ABABAB… formation. Nylon 6 is an exception in that it is made from only one monomer, which is a cyclic compound.

a) Draw the structure of the repeat unit a polyamide that is made from benzene-1,4-dicarboxylic acid and benzene-1,4-diamine. **[3 marks]**

b) Draw a diagram showing how hydrogen bonding could arise between the chains of this polymer. **[1 mark]**

c) The molecular formula for the monomer used to make nylon 6 is C_5H_9NO. Suggest a possible structure for this monomer. **[3 marks]**

d) Write an equation for the polymerisation reaction that forms nylon 6. **[1 mark]**

e) How is this different from other polyamide polymerisations? **[1 mark]**

4. The diagram below shows a length of the polymer PET.

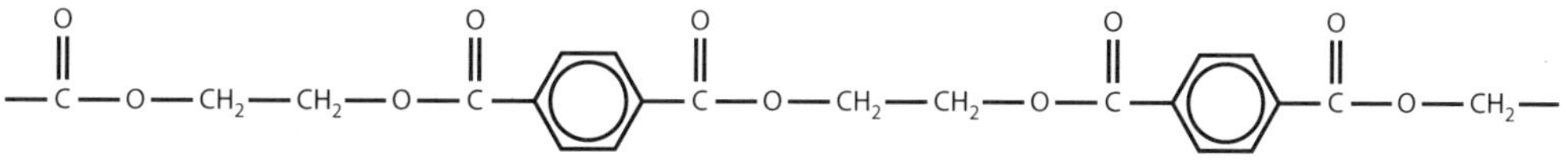

a) Draw the structure of the two monomers from which PET is made. **[2 marks]**

b) What type of polymer is PET? **[1 mark]**

c) This type of polymer is more easily broken down into its monomer units than many other types of polymer. Explain why and describe some advantages of this property of PET. **[4 marks]**

Organic Applications

Esters

Esters form the basis of many flavours and aromas: Their comparatively low boiling point means that they are volatile and vaporise at room temperature, and interact with smell receptors in the nose.

Methyl methanoate
raspberry

3-methylbutylethanoate
banana

Esters make good organic solvents: Only small esters are water soluble but larger esters make good organic solvents. The carbonyl group means that they are able to bond with polar regions of organic molecules.

Esters are used as plasticisers: Some polymers can be made more flexible by adding a plasticiser. This moves the polymer chains apart and allows them to slide over each other more easily.

The ester di-(2-ethylhexyl)hexanedioate is used to make PVC flexible.

Oils and Fats are Esters

Naturally occurring fats and oils are triesters of propan-1,2,3-triol with long chain carboxylic acids. They are commonly known as **triglycerides**.

The three carboxylic acids forming the triester are typically different from each other and may be saturated or unsaturated. Long chain carboxylic acids are known as **fatty acids**.

Some common fatty acids	Common name
$CH_3(CH_2)_{14}COOH$	Palmitic acid
$CH_3(CH_2)_{16}COOH$	Stearic acid
$CH_3(CH_2)_7CH{=}CH(CH_2)_7COOH$	Oleic acid

Naturally occurring unsaturated fatty acids are the cis isomer. This introduces a bend in the molecule. The bend interferes with intermolecular bonding and lowers the melting point of the molecule.

Natural fats contain a mixture of different triesters. The proportion of each triester is characteristic of the source material.

Fats are solid at room temperature and tend to have a lower percentage of unsaturated fatty acids than oils, which are liquids at room temperature.

Fats Can Be Hydrolysed to Make Soaps

The salts of long chain carboxylic acids are known as soaps. Heating fats with alkali produces soaps and propan-1,2,3-triol.

NaOH(aq)

Biodiesel is Made by Transesterification of Oils

Biodiesel is a mixture of the methyl esters of long chain carboxylic acids. The methyl ester is prepared using oils, alkali and methanol.

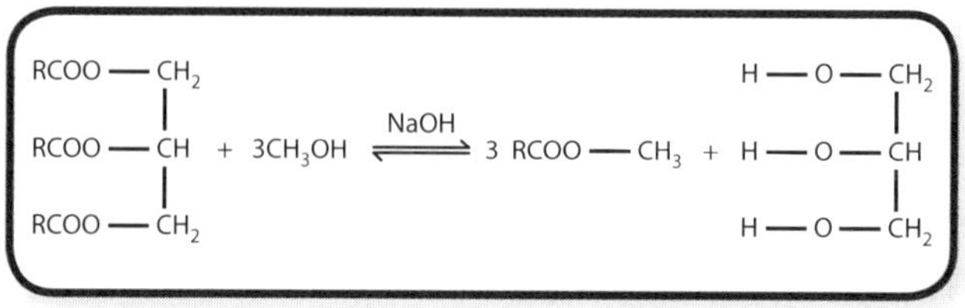

$$\begin{array}{l} RCOO-CH_2 \\ \quad\quad\quad\;| \\ RCOO-CH \\ \quad\quad\quad\;| \\ RCOO-CH_2 \end{array} + 3CH_3OH \overset{NaOH}{\rightleftharpoons} 3\ RCOO-CH_3 + \begin{array}{l} H-O-CH_2 \\ \quad\quad\quad\;\;| \\ H-O-CH \\ \quad\quad\quad\;\;| \\ H-O-CH_2 \end{array}$$

The alkali acts as a catalyst in the reaction. Since the reaction is at equilibrium, excess methanol is used to keep the position of equilibrium to the right. The excess methanol and catalyst can be recycled to use in the next reaction.

Dyes

Dyes are substances that can be incorporated into the bulk of a material to give it colour.

Molecules are coloured because they absorb some wavelengths of light from the visible spectrum and reflect others (see azo dyes on page 82). As the frequency of light absorbed changes, so the colour changes.

Organic molecules that are coloured have **extended delocalised electron systems**.

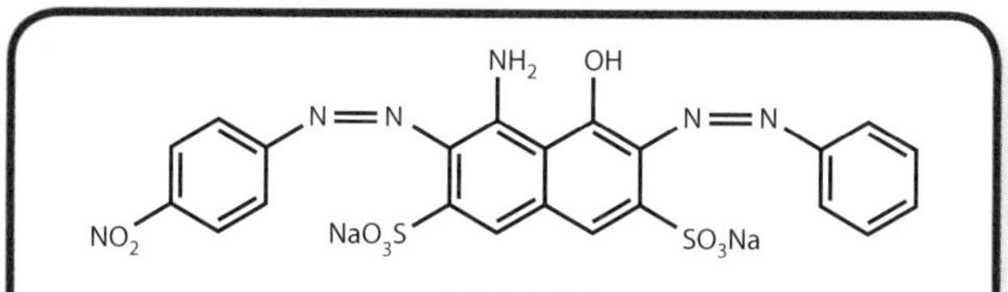

Acid black dye

The part of the molecules that is responsible for the colour of the molecule is known as the **chromophore.** Substituents may be attached to the chromophore in order to modify the colour, increase the solubility or improve the ability of the dye to bind to the fibre.

How Dyes Attach to Cloth Fibres

Dyes can attach to cloth fibres in a number of ways.

1. Intermolecular bonding. Many dyes are able to hydrogen bond to groups on the fibre. For example cotton fibres contain many —OH groups that can hydrogen bond with groups in the dye. For effective dyeing the molecules must have several attachment points to the fibre to counterbalance the fact that each bond is weak. These are known as **direct dyes**.

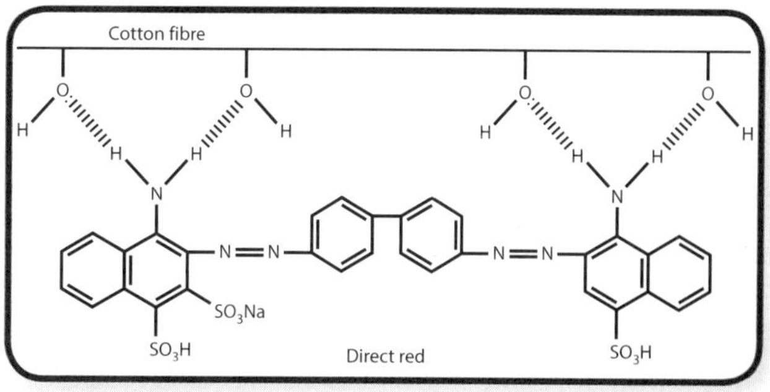

Direct red

2. Ionic bonding. Wool and silk fibres are proteins that have ionisable $—NH_2$ and —COOH groups in their R side groups (see page 84). Dyes that have ionised groups such as $—SO_3^-$ will bond to the fibres by electrostatic attraction. These are known as **acid and cationic dyes**. They are typically used for wool and polyamides.

3. Covalent bonding. Reactive dyes contain a group that will form a covalent bond with the fibre. These are known as **fibre reactive dyes.** A common method is to attach trichlorotriazine to the chromophore of the dye.

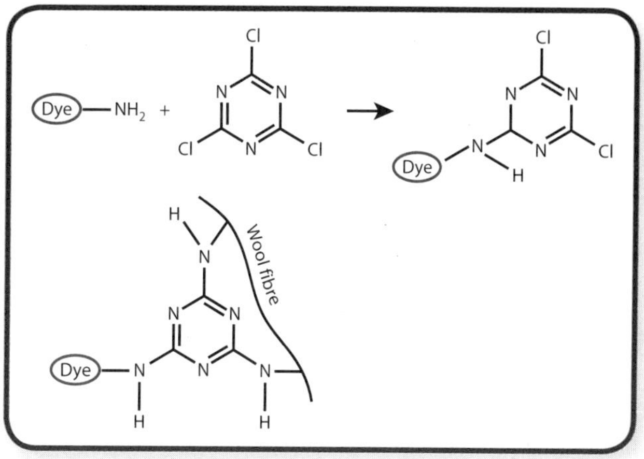

The dye substitutes for one Cl atom on the triazine ring. NH_2 groups on the fibre then substitute for the other two Cl atoms on the ring.

4. Precipitation within the fabric fibres. Cloth such as denim is soaked in the colourless, soluble form of indigo. An oxidising agent is added that converts the indigo to the reduced, insoluble blue form. These are known as **vat dyes**.
5. Coordinate bonding. These dyes are added to the fabric together with a metal ion such as aluminium or chromium, known as a **mordent**. The metal ion forms a complex that includes both the dye and the fibre as ligands and so binds the dye to the cloth.

Making Dyes Soluble

Adding ionic groups or groups that can hydrogen bond is likely to increase the solubility of a dye. Common groups that are added to the chromophore to increase solubility are —SO_3Na, the salt of sulfonic acid, or —COONa, the salt of a carboxylic acid.

Adding Substituents to the Chromophore Changes the Colour

Substituents groups that are able to donate or withdraw electrons from the chromophore are most likely to change the colour, e.g. —NH_2, —OH, NO_2.

C_6H_5—N=N—C_6H_4—NH_2

Yellow

O_2N—C_6H_4—N=N—C_6H_4—$N(C_2H_5)_2$

Red

The chromophore contains electrons that have small differences in the energy between their ground state and excited state. They are able to absorb the energy of photons in the visible spectrum and become excited.

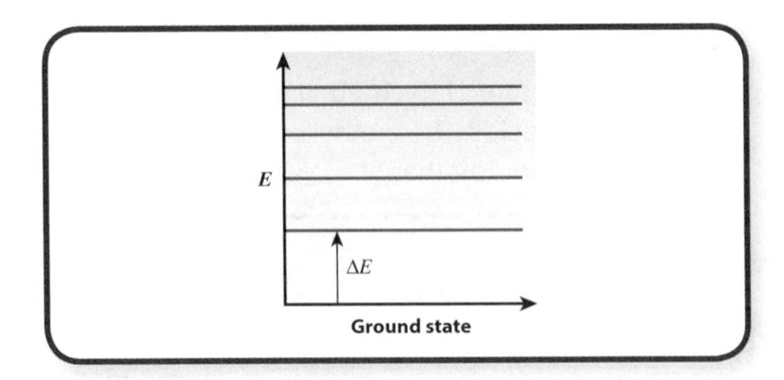

The energy of the photon absorbed = ΔE, which is the difference between the ground state and the excited state. The frequency of light from which the photon is absorbed is given by $E = hf$ where h = Planck's constant and f = frequency of light.

The remaining frequencies of visible light are reflected and produce the colour seen. This colour is complementary to the frequency of light absorbed.

Changing substituents on the chromophore changes the value of ΔE and hence the colour.

Note: You will not be expected to know any of the dye structures but should be able to recognise the chromophore and groups that increase solubility.

QUICK TEST

1. What is a plasticiser?
2. Name the functional group in a triglyceride.
3. What is transesterification?
4. How are soaps made?
5. How is trichlorotriazine used in the dye industry?
6. What group could be added to a dye to increase solubility?

PRACTICE QUESTIONS

1. Which statement is **not** true of naturally occurring fats? **[1 mark]**

A They can be hydrolysed by alkali.

B Fats from a single source contain only one type of fatty acid.

C They may contain carbon–carbon double bonds.

D They are esters.

2. Which type of dye makes the strongest bond to fabrics? **[1 mark]**

A Direct dyes

B Vat dyes

C Fibre reactive dyes

D Cationic dyes

3. Fats and oils are triesters of propan-1,2,3-triol with long chain carboxylic acids, which may be saturated or unsaturated.

a) Draw the structure of a triester of propan-1,2,3-triol. Use R—COOH as the formula of the carboxylic acids. **[2 marks]**

Naturally occurring unsaturated fatty acids are always in the cis (Z) configuration.

b) Explain the effect this has on the physical properties of unsaturated fats. **[3 marks]**

Soap has been make in the UK since the 13th century. It was originally made by boiling fats with lye, a solution made by mixing wood ash and rainwater.

c) Suggest what is present in lye that causes fats to be converted into soaps. **[1 mark]**

d) Write an equation to show the reaction that converts fats to soaps using your structural formula from part **a)**. **[2 marks]**

4. The diagram shows the structure of the dye direct blue 53.

a) Suggest how this group could attach to cotton. **[2 marks]**

b) Which groups in the dye increase the solubility? **[1 mark]**

c) Circle the chromophore in this molecule. **[1 mark]**

d) Explain what gives rise to colour in organic molecules. **[4 marks]**

e) Direct blue 53 is one of a group of direct blue dyes with very similar structure but different colours. Explain why a very small modification to the structure produces a different colour. **[2 marks]**

Chromatography

Chromatography is a technique used to separate out the components of a mixture on the basis of their division between two substances in different phases. The two phases may be a liquid and a solid or a liquid and a gas.

One of the phases remains stationary and the other, mobile phase, moves over the stationary phase. The components are separated because of their different attractions to the two phases.

The principles of all types of chromatography can be illustrated by column chromatography.

Column Chromatography

In simple column chromatography a glass column with a tap at the bottom is packed with tiny beads of silica gel. This is the stationary phase. The column is filled with solvent (eluent), the mobile phase.

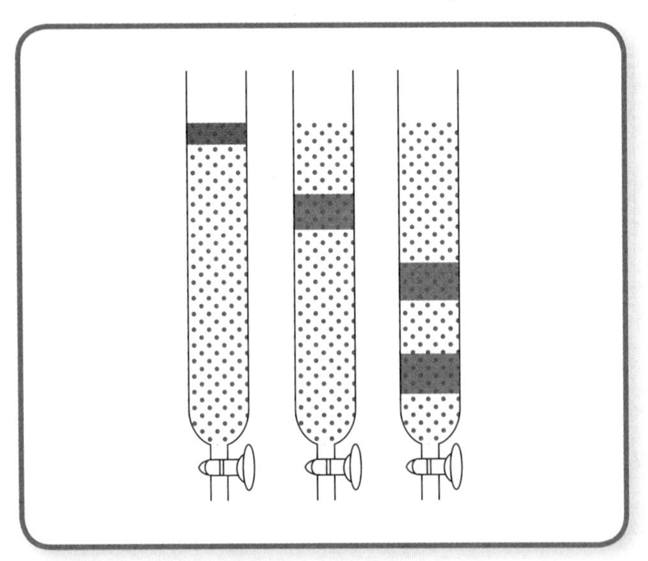

The mixture to be separated is placed on top of the gel. The tap is opened, allowing the solvent and mixture to run slowly through the column, while the column is continually topped up with more solvent.

One component of the mixture is attracted to the gel and is continually adsorbed to the surface of the beads and then washed off by fresh solvent coming past it. This adsorption slows down the rate at which it moves down the column.

The other component of the mixture dissolves well in the solvent and has less tendency to adsorb onto the gel. It moves rapidly down the column with the solvent.

The further the mixture travels through the column, the bigger the separation.

The difference in progress down the column means that the two components leave the column at a different volume of solvent and can be collected separately as they emerge.

High Performance Liquid Chromatography

HPLC is a faster, more effective type of column chromatography. The solvent is pumped through the column under high pressure and the beads are very small with a large surface area. This gives extremely good separation of components.

The nature of the beads and the solvent are chosen to give a good separation for the mixture under study.

The time taken for a substance to pass through the column is called the **retention time**. Each substance has a consistent retention time for a given column, solvent and pressure.

The column can be calibrated by passing known substances through it and measuring their retention times.

UV absorption is often used to detect the components as they emerge from the column.

The HPLC can be linked directly to a mass spectrometer so that a mass spectrum can be made of each component as it emerges from the column.

Gas–Liquid Chromatography

GLC or GC (gas chromatography) uses a liquid stationary phase and a gaseous mobile phase.

The liquid used is a high boiling point solvent, which is coated onto inert particles and packed into a long metal column. This provides a very large surface area of liquid.

An inert gas such as nitrogen is pushed through the column under pressure. This is known as the **carrier gas**.

The volatile sample is injected into the column and is carried through by the gas. As it passes over the liquid it continually dissolves into the liquid and vaporises again.

Substances that vaporise readily and are less soluble in the liquid travel through the column quickly and have short retention times.

Substances that are less volatile and more soluble in the liquid have longer retention times.

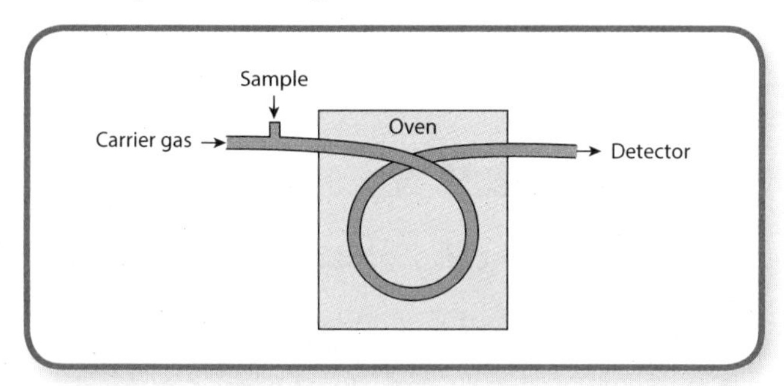

The column is inside a thermostatically controlled oven. This keeps the retention time the same for each run.

A recorder gives a trace where the quantity of the substance is proportional to the area under the peak. If the components in the mixture are known then the quantity can be calculated. A set of standard solutions of different known concentrations is run through the GLC to find the exact relationship between peak area and concentration. This is known as an **external standard**.

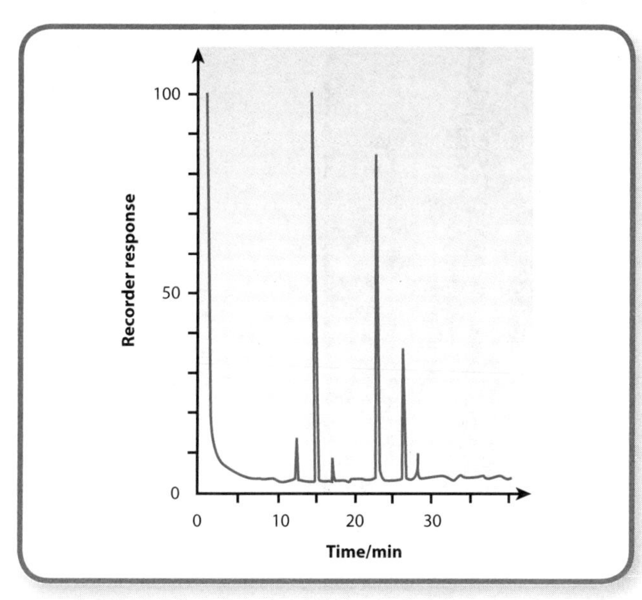

For molecules of the same type, the lower the molecular mass the shorter the retention time.

If the components in the mixture are unknown, the GLC can be linked to a mass spectrometer so the compounds can be analysed as they emerge from the column.

GLC is a highly sensitive technique that can detect very small amounts. This makes it suitable for use when the sample size is limited, such as in forensics or drug testing.

GLC has limitations because it is only suitable for volatile substances. It is often necessary to chemically convert the sample into something more volatile before it can be used.

Unknown substances may have very similar retention times so peaks may overlap or contain more than one component.

Thin Layer Chromatography

The stationary phase is a thin layer of very fine particles of a solid such as silica gel on a rigid support plate. The mobile phase is a solvent.

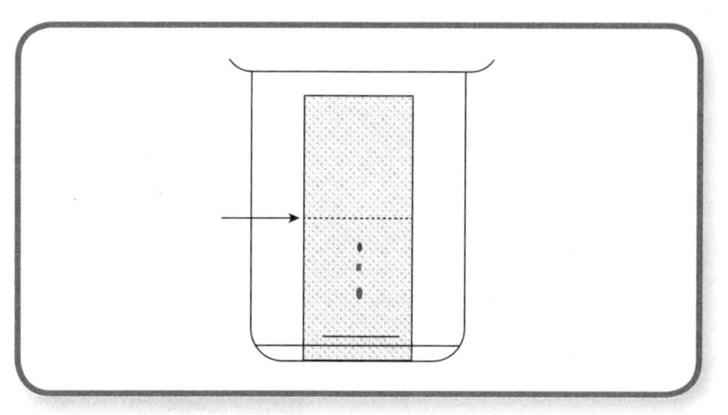

- The test substance is dissolved and spotted onto the thin layer plate about 1 cm from the bottom using a capillary tube. The spot is kept to as small an area as possible.
- The plate is placed in a developing tank containing solvent that is below the level of the sample spot and covered. The tank is covered to allow a saturated atmosphere of solvent to form.
- The solvent is allowed to rise up the plate by capillary action but there is no net evaporation because of the saturated atmosphere in the tank.
- When the solvent is near the top the position of the solvent front is marked.

- The plate is removed and dried. If the sample is not coloured, a locating agent such as iodine vapour can be used, which stains the component spots. The plate is placed in a beaker containing a few crystals of iodine.

The location of the spots in the mixture is described by the R_f value. This is the ratio of the distance moved by the component to the distance moved by the solvent.

$$R_f = \frac{\text{distance moved by spot}}{\text{distance moved by solvent front}}$$

The identity of the components of the mixture can be found by spotting known standards alongside the mixture.

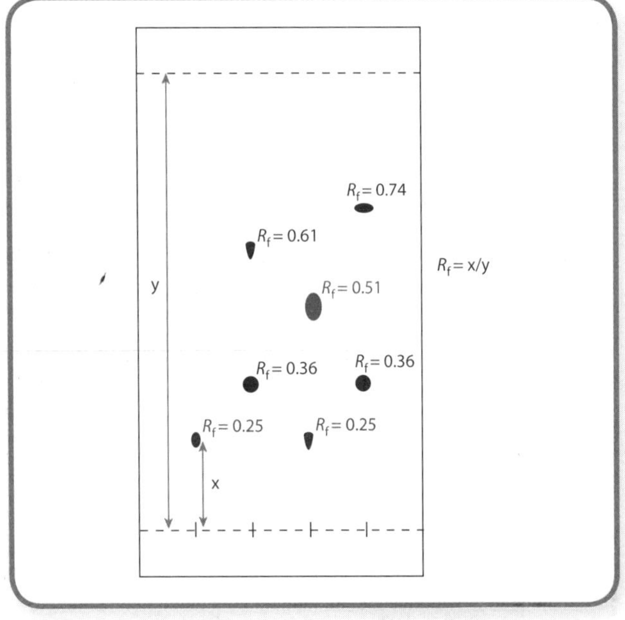

Paper Chromatography

Paper chromatography is similar to TLC but a special absorbent paper is used in place of the TLC plate.

The mixture and standards are spotted onto the bottom of the paper, which is suspended in the solvent in a saturated atmosphere of solvent.

When the solvent has reached the top of the paper the chromatogram is removed and allowed to dry.

The spots can be located with a specific chemical reaction. For example, to detect amino acids the dried paper is sprayed with ninhydrin and heated. Amino acids react with the ninhydrin to give a purple colour.

QUICK TEST

1. What is the mobile phase in HPLC?
2. What is the stationary phase in GLC?
3. On what basis do the components of a mixture separate in GLC?
4. What is meant by an 'external standard' in GLC?
5. Why is the lid of a chromatography tank kept covered when a chromatogram is running?
6. How is the R_f value calculated?

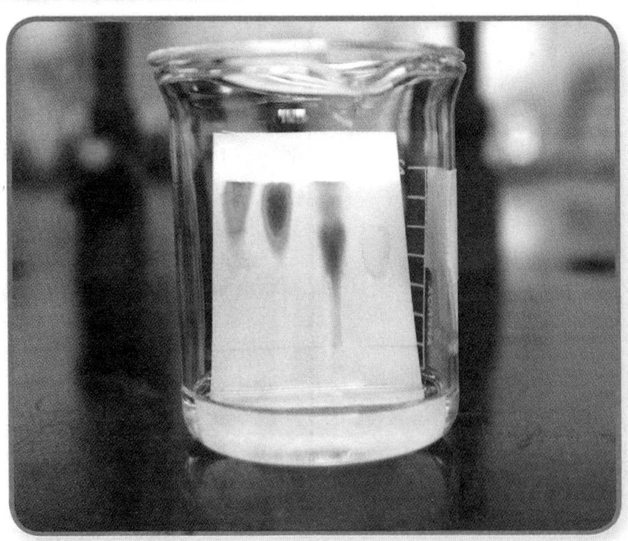

PRACTICE QUESTIONS

1. In a TLC plate the solvent front travelled 7.5 cm. A standard spot of aspirin travelled 5.4 cm and an organic preparation of aspirin showed spots at 1.8 cm and 5.4 cm from the origin. Which is true? **[1 mark]**

   A The preparation did not contain aspirin. ☐

   B The R_f value of aspirin in this system is 1.38. ☐

   C The R_f value of a contaminant in the aspirin preparation is 0.24. ☐

   D A different solvent is needed to separate the components. ☐

2. Which of the following is **not** a concern with GLC? **[1 mark]**

   A The sample must be volatile. ☐

   B The components of the mixture must have sufficiently different retention times. ☐

   C The components must be readily detectable by a chemical test. ☐

   D The machine must be calibrated for each component. ☐

3. The gas–liquid chromatogram below show the results of an analysis of flavour compounds in a sample of French dressing.

   a) State the relationship between molecular mass and retention time in this sample. **[1 mark]**

   b) How was it possible to identify which peak represented each acid as labelled? **[2 marks]**

   c) The label on the dressing claimed it contained no more than the permitted level of 150 mg dm^{-3} of benzoic acid. How could this chromatogram be used to confirm this claim? **[3 marks]**

   d) The longer chain acids cannot be analysed directly using GLC but must first be converted to their methyl esters. Explain why. **[3 marks]**

   e) Unknown compounds can be separated out from mixtures using GLC. What method is often used to analyse the structure of these components? **[1 mark]**

4. Paper chromatography is a simple technique that can be used to separate components of a mixture.

   Describe how you could use paper chromatography to identify the amino acids in a hydrolysed dipeptide. **[5 marks]**

Carbon-13 NMR Spectroscopy/ Mass Spectrometry

Carbon-13 NMR Spectroscopy

NMR spectroscopy measures energy released from the nucleus of the atoms of molecules as a result of absorbing radio frequency electromagnetic energy.

Carbon-13 spectroscopy measures the signals coming from the nuclei of carbon-13 atoms. C-13 is an isotope of carbon with an abundance of approximately 1/100 carbon atoms.

The sample is placed in a strong magnetic field. A spectrum of radio frequency electromagnetic radiation is passed through it. Some specific wavelengths are absorbed by C-13 nuclei. The absorbed energy is then released and recorded on a spectrum.

The exact wavelength of energy absorbed and released depends on the position of the carbon atom in the molecule – the **carbon environment**. The spectrum shows the difference between the carbon giving the signal and a standard carbon in TMS.

The standard used is tetramethylsilane (TMS). The peak produced by this is set as zero (it is not always shown as a peak on the spectrum). The difference between the TMS signal and the peak is called the chemical shift (δ), measured in parts per million.

Example spectrum: Propan-1-3-diol

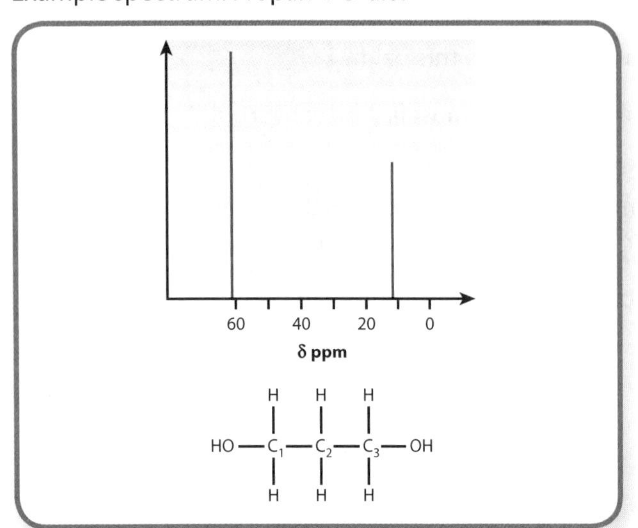

In a C-13 spectrum, only the carbons atoms give a signal.

The Number of Carbon Environments

There are two carbon environments giving two signals in the spectrum.

Carbons 1 and 3 are both attached to two hydrogen atoms, one —OH group and one $—CH_2$ group. They are in identical environments and give the same signal. If molecules have a plane of symmetry, the carbons on each side are in the same environment.

Carbon 2 is attached to two $—CH_2OH$ groups. This is a unique carbon environment for this molecule and gives a different signal.

The Shift Value of Carbon in Different Environments

The relevant typical shift values for different carbon environments will be provided in the exam.

Table of values

Type of carbon	Chemical shift, δ/ppm
C—C	5–55
C—Cl or C—Br	30–70
C—N (amines)	35–60
C—O	50–70
C=C (alkenes)	115–140
aromatic	110–165
carbonyl (ester, carboxylic acid, amide)	160–185
carbonyl (aldehyde, ketone)	190–220

The shift value for carbon 2 is in the normal alkane range. The shift value for carbons 1 and 3 is higher. Electronegative atoms bonded to the carbon atom tend to increase the shift value.

Example spectrum: Propan-1,2-diol

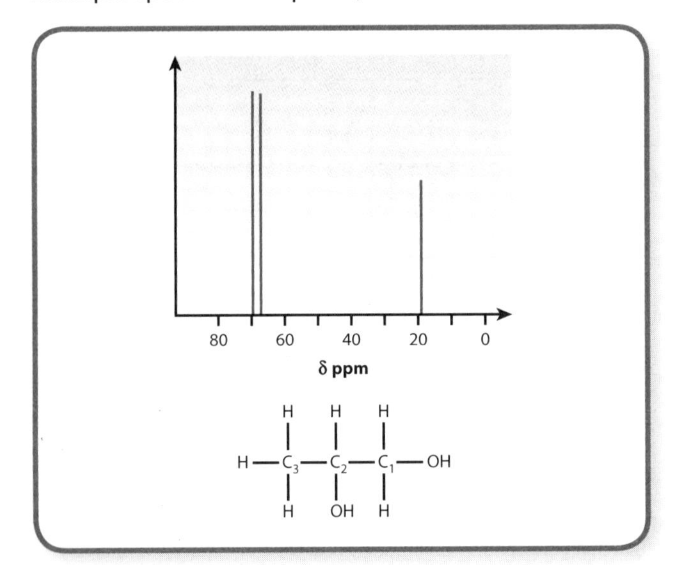

This time each C has a different environment giving three signals.

C_1 is bonded to 2 × H and 1 × OH

C_2 is bonded to 1 × H, 1 × CH_3, 1 × CH_2OH and 1 × OH

C_3 is bonded to 3 × H and 1 × $CH(OH)CH_2OH$

Don't forget...

In a carbon-13 spectrum:

- **the number of signals shows the number of carbon environments**
- **the shift value of the signal shows the type of environment**
- **the relative heights of the signals give no information for A-Level.**

Eliminating Possible Structures on the Basis of a C-NMR Spectrum

Chemical analysis, IR spectroscopy and mass spectrometry may provide information about the empirical formula, functional groups and M_r of an unknown compound. This often leaves a number of possible isomers. The C-NMR spectrum may help to eliminate some isomers.

e.g. IR indicates a ketone, mass spectrometry indicates an M_r of 86.

Possible isomers:

5 carbon environments

3 carbon environments

4 carbon environments

Carbon-13 NMR spectrum:

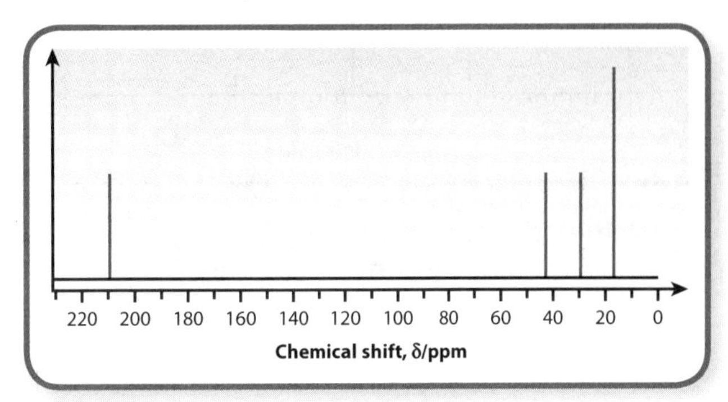

The spectrum shows four carbon environments therefore the molecule is 3-methylbutan-2-one.

Carbon Environments in an Aromatic Compound

Carbons that are symmetrical within a benzene ring or a cyclic compound are in the same environment.

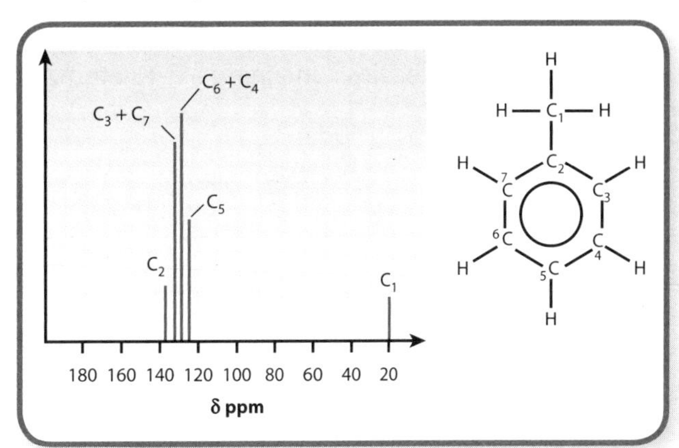

Predicting a Spectrum from the Molecular Structure

Propan-2-ol:

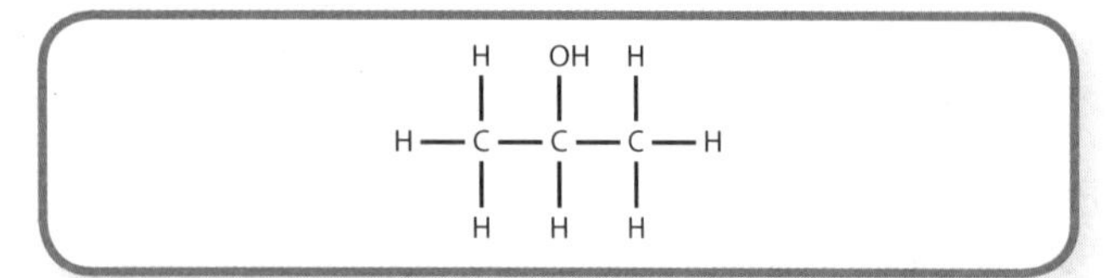

There are two carbon environments.

CH_3 has a shift in the range 20–40 ppm.

CH(OH) has a shift in the range 50–70 ppm.

Height of signal is not significant.

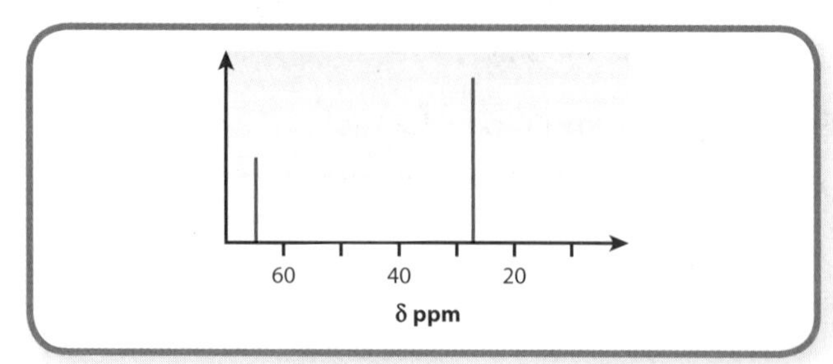

Mass Spectrometry

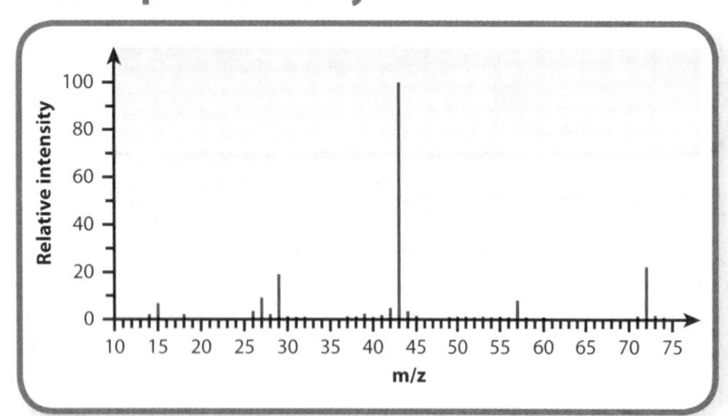

The spectrum shows positive ions that have been produced by fragmenting the whole molecule.

The peak with the highest m/z value comes from the molecular ion and gives the mass of the whole molecule.

The difference between the m/z value for the molecular ion peak and the other peaks is the mass of fragments that have been broken off the whole molecule.

The most abundant fragment peak is set at 100% and is known as the base peak.

A tiny peak at m/z one higher than the molecular ion peak is caused by molecules containing a C-13 atom.

Common fragment ions include:

m/z	Possible ion
15	$[CH_3]^+$
29	$[CH_3CH_2]^+$
43	$[CH_3CH_2\ CH_2]^+$ $[CH_3CO]^+$
77	$[C_6H_5]^+$

For the spectrum shown:

$M_r = 72$ Base peak = m/z 43

Possible identity of peak at 29 = $[CH_3CH_2]^+$

Possible identity of peak at 15 = $[CH_3]^+$

The spectrum is of butanone.

$M_r = 72$

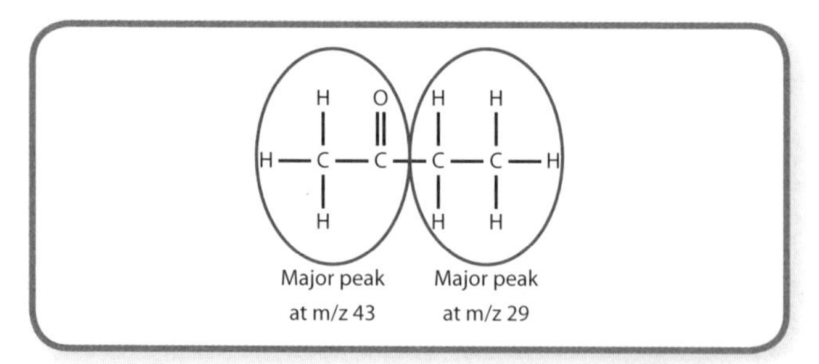

QUICK TEST

1. What information is given by the chemical shift value in a carbon-13 NMR spectrum?
2. How many carbon environments are there in butane?
3. What information is given by the peak height in a carbon-13 NMR spectrum?
4. How can C-NMR spectra help distinguish between different structural isomers?
5. What gives rise to peaks in a mass spectrum?
6. Suggest a likely assignation of a peak at m/z 77.

PRACTICE QUESTIONS

1. The carbon-13 spectrum of pentan-3-one would show **[1 mark]**

   A three peaks

   B a peak at 160–180

   C one peak smaller than the others from the C=O signal

   D five peaks.

2. The mass spectrum of propanone **[1 mark]**

   A will only contain three peaks

   B will have a peak at 58 due to C_3H_6O

   C is likely to have a large +1 peak

   D is likely to show major peaks at 15, 43 and 58.

3. The structures shown are all isomers of C_9H_{12}.

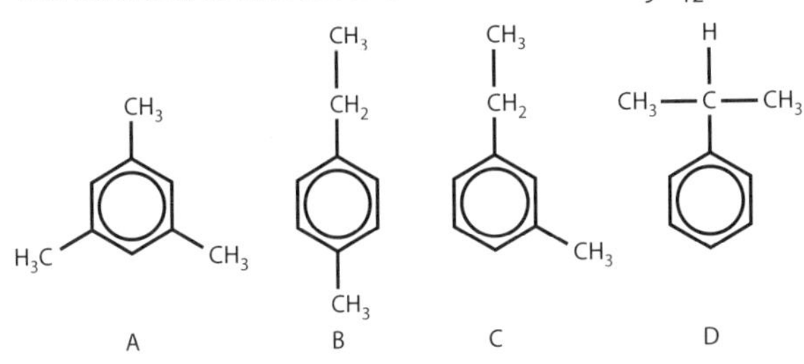

   a) Sketch the C-13 spectrum that you would expect to be produced by isomer A. **[2 marks]**

   The spectrum below shows the C-13 of one of the three isomers B, C or D.

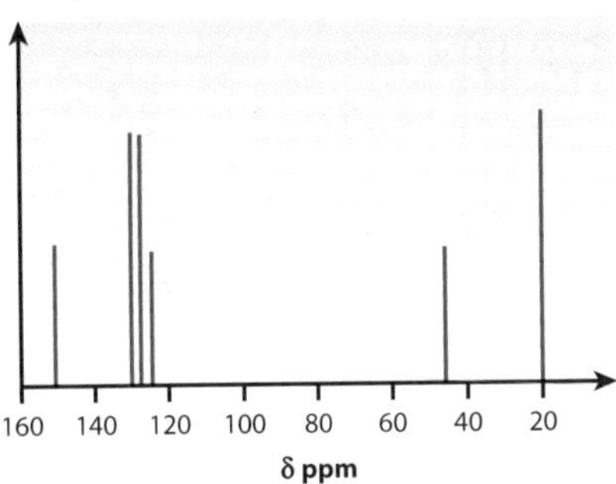

   b) Which isomer produced this spectrum? Explain your answer. **[2 marks]**

4. This mass spectrum represents a molecule that fizzes when added to sodium carbonate.

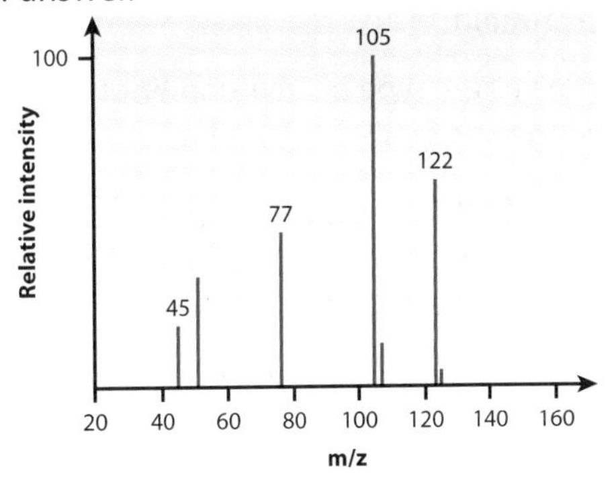

   a) What is the molecular mass of this molecule? **[1 mark]**

   b) Suggest the fragment that gives a peak at m/z 77. **[1 mark]**

   c) What fragment might have been lost from the molecular ion to give the peak at 77? **[1 mark]**

   d) Suggest a structure for this molecule. **[1 mark]**

Proton NMR Spectroscopy

Hydrogen atoms give a signal in a proton NMR spectrum. There is no signal from any other atom.

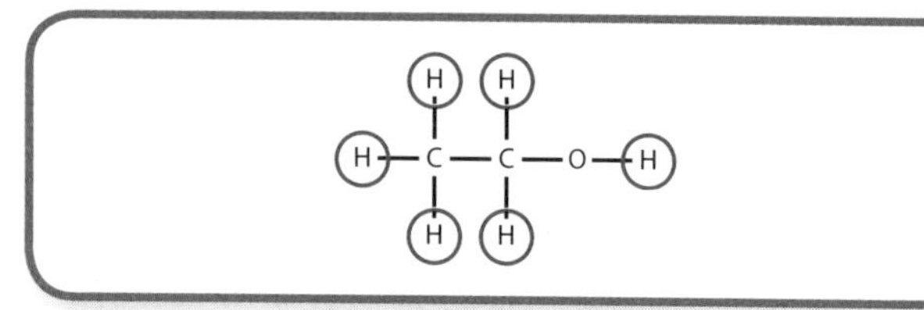

Proton signals from the solvent would confuse the spectrum. Deuterium (^2H) gives no signal, so deuterated solvents such as $CDCl_3$ are used to dissolve the sample.

The Number of Hydrogen Environments

The number of peaks in the spectrum = number of hydrogen environments.

The hydrogen environment is the types of group bonded to the same carbon atom as the hydrogen giving the signal.

Hydrogens attached to the same carbon atom give the same signal.

Hydrogens that are symmetrical in the molecule give the same signal.

3 hydrogen environments 2 hydrogen environments

The Type of Environment

The chemical shift gives the type of environment for the hydrogens giving the signal.

Tetramethylsilane (TMS), $(CH_3)_4Si$, is used as the standard to set 0, as in C-13 NMR (see page 104). The protons in TMS are very shielded so protons from other compounds will not absorb at a lower value.

TMS contains 12 protons in the same environment and so gives a strong signal.

Type of proton	Chemical shift, δ/ppm
R—CH_3	0.7–1.6
R—CH_2—OR	1.2–1.4
H_3C—C(=O)— R—CH_2—C(=O)—	2.0–2.9
C_6H_5—CH_3 C_6H_5—CH_2—R	2.3–2.7
X—CH_3 X—CH_2—R (X = halogen)	3.2–3.7
—O—CH_3 —O—CH_2—R	3.3–4.3
R—O—H	3.5–5.5
C_6H_5—H	7
H_3C—C(=O)—H	9.5–10
H_3C—C(=O)—OH	11.0–11.7

Note: The shift values are very approximate; they can and do vary from the ranges shown.

δ 0.7–1 ppm δ 3.3–4.3 ppm δ 3.5–5.5 ppm

Identifying O—H and N—H Protons

O—H and N—H protons can be hard to identify because they can absorb over a wide range of shift values. When D_2O is added to the solvent the O—H and N—H protons exchange for D and the peak disappears. Running a second spectrum with D_2O helps to identify these peaks.

The Ratio of Hydrogens in Each Environment

The area under the peaks in the spectrum shows the ratio of hydrogen atoms in the different peaks.

This may be given as a number above the peak or as an integration trace.

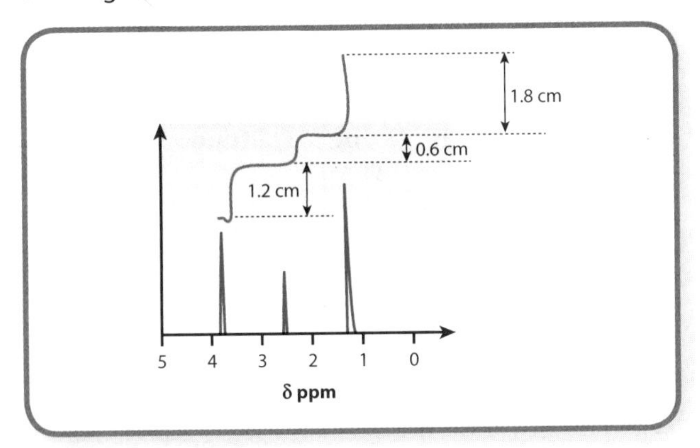

Measure the vertical sections of the integration trace and convert into a whole number ratio.

1.8 : 1.2 : 0.6 = 3 : 2 : 1

The integration trace gives the ratio of hydrogens not the absolute number so 3 : 2 : 1 could be 6 : 4 : 2.

The Number of Hydrogens on Adjacent Carbons

The signal from one hydrogen environment is split by any hydrogens on the adjacent carbon.

The peak is split into a number of peaks equal to the number of hydrogen atoms on the adjacent carbon atom + 1 (the n+1 rule).

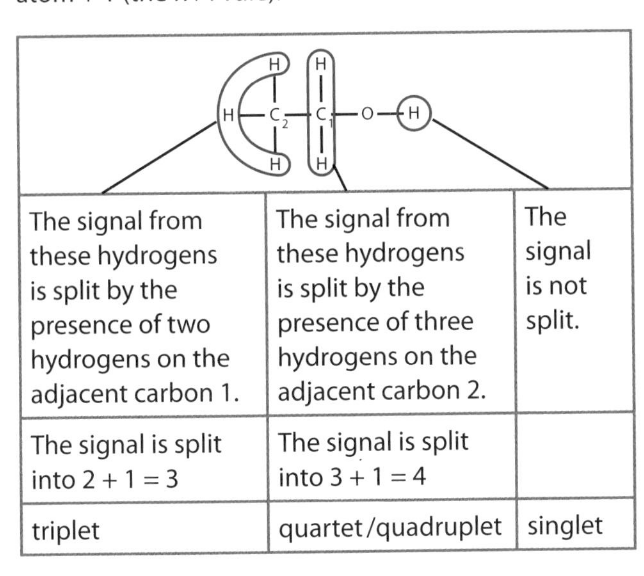

The signal from these hydrogens is split by the presence of two hydrogens on the adjacent carbon 1.	The signal from these hydrogens is split by the presence of three hydrogens on the adjacent carbon 2.	The signal is not split.
The signal is split into 2 + 1 = 3	The signal is split into 3 + 1 = 4	
triplet	quartet/quadruplet	singlet

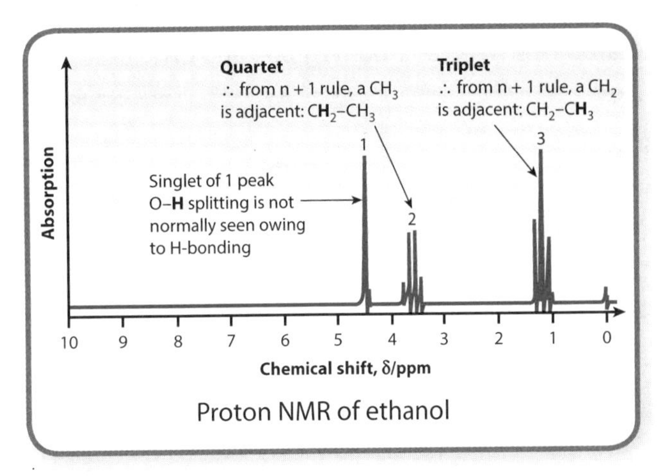

Proton NMR of ethanol

To identify a structure from a proton NMR spectrum: Draw up a table showing all the information from the spectrum.

Example 1: The spectrum of an ester is shown below.

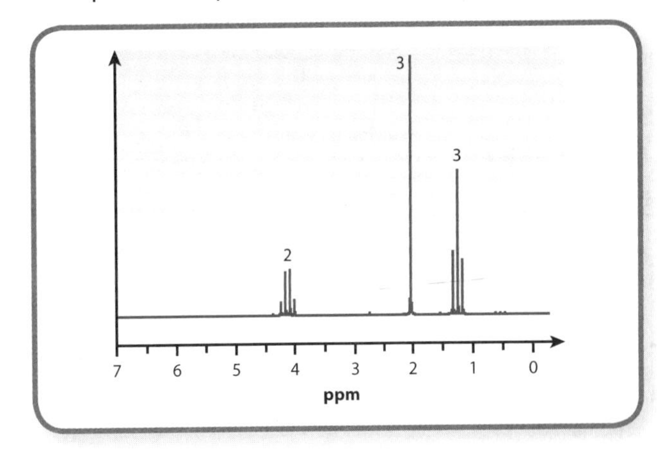

Shift/ ppm	Environment	No. of H	Splitting	Adjacent protons
1.2	R—C—**H**	3	triplet	2
2.0	O—C—**H**	3	singlet	0
4.2	O=C—C—**H**	2	quartet	3

The peak at 2.0 ppm suggests three protons on a carbon atom bonded to an O.

H
|
H—C—O—
|
H

The peak at 4.2 ppm suggests two hydrogen atoms on a carbon between a carbonyl group and CH_3.

—C(=O)—CH₂—CH₃

The peak at 1.2 ppm suggests three hydrogen atoms on a carbon next to CH_2.

—CH_2—C(H)(H)—H

Knowing the compound is an ester allows all the information to be put together.

H—C(H)(H)—O—C(=O)—C(H)(H)—C(H)(H)—H

Methylpropanoate

Example 2: A compound with molecular formula C_8H_{10}

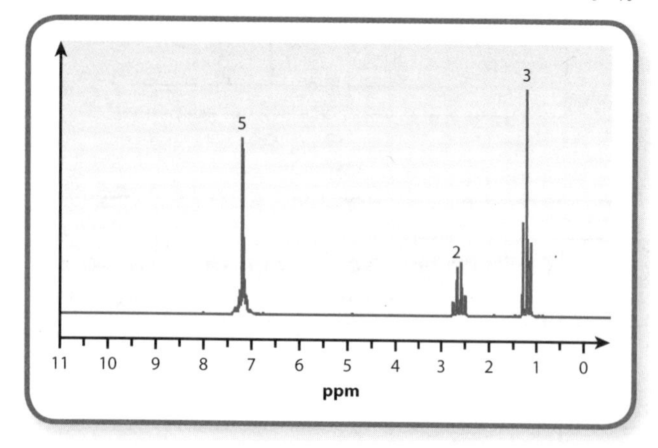

The spectrum does not always give the area under the peak or the integration trace.

The protons in a benzene ring can split with each other to give a complex pattern of peaks that cannot always be resolved into individual peaks.

This is known as a multiplet.

The peak at 7.2 ppm is split into many overlapping peaks. A multiplet at a shift value of around 7 ppm is likely to be a benzene ring.

Shift/ ppm	Environment	No. of H	Splitting	Adjacent protons
1.2	R—C—**H**	3	triplet	2
2.6	(benzene ring)—C(**H**)(**H**)—	3	quadruplet/ quartet	3
7.2	(benzene ring)—**H**	5	multiplet	

(benzene ring)—C(H)(H)—C(H)(H)—H

Phenylethane

QUICK TEST

1. Why does proton NMR use a deuterated solvent?
2. What is TMS?
3. How can the number of proton environments be determined?
4. How can the number of protons in a particular environment be determined?
5. How can the type of proton environment be decided?
6. What can be deduced from the fact that a signal is split into a quartet?

PRACTICE QUESTIONS

1. TMS is chosen as a standard for proton NMR spectroscopy. **One** reason is because **[1 mark]**

   A it has no protons to confuse the signal ☐

   B it will swap with the protons of —OH groups ☐

   C it contains a silicon atom ☐

   D it has very shielded protons. ☐

2. The number of peaks on the proton NMR spectrum of methylpropan-2-ol is  **[1 mark]**

   A 4 ☐

   B 3 ☐

   C 2 ☐

   D 1 ☐

3. 3-methylbutan-2-one is shown below.

```
     H   H   O   H
     |   |   ||  |
  H—C—C—C—C—H
     |   |       |
     H   |       H
      H—C—H
         |
         H
```

   a) How many peaks will this molecule produce in its proton NMR spectrum? **[1 mark]**

   b) The hydrogen on carbon 3 gives a signal that is split into a multiplet with seven peaks Explain why. **[1 mark]**

   c) The signal from the hydrogens on carbon 1 has an area three times the size of the signal from that on carbon 3. Explain why. **[1 mark]**

   d) Suggest the shift value of the signal from the hydrogens on carbon 1. **[1 mark]**

   e) Sketch a diagram of the spectrum you would expect to see. **[4 marks]**

4. The diagram shows the proton NMR spectrum for an ester with the molecular mass 88.

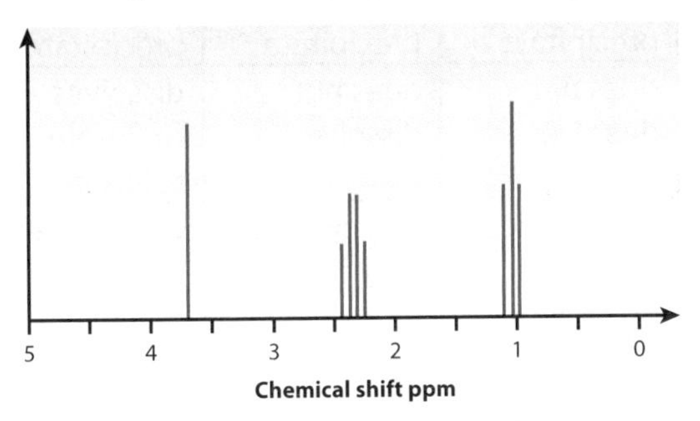

   Suggest a structure for this molecule and show your reasoning. **[5 marks]**

DAY 7 60 Minutes

Summary of Reactions/ Organic Pathways

Summary of Reactions

Test Tube Reactions for Functional Groups

Functional group	Test	Results
alkene	Shake with bromine water.	Colour change from brown to colourless
alcohol (general)	Add PCl_5 to dry sample.	Steamy fumes (HCl) form. Note: Also reacts with carboxylic acids.
distinguishing between 1°, 2° and 3° alcohols	Warm with acidified potassium dichromate.	Colour change from orange to green for 1° and 2° alcohols No colour change for 3° alcohols
carbonyl groups in ketones and aldehydes	Add 2,4-DNP in methanol with H_2SO_4 (Brady's reagent).	Orange/yellow precipitate
distinguishing between aldehydes and ketones	Warm with ammoniacal silver nitrate (Tollens' reagent). **or** Heat with Fehling's solution.	Silver mirror forms with aldehydes. No change with ketones. Brick red precipitate forms with aldehydes. No change with ketones.
carboxylic acid	Add sodium carbonate solution.	Effervescence
haloalkane	Warm with aqueous sodium hydroxide. Neutralise and add silver nitrate solution.	White (Cl), cream (Br) or yellow (I) precipitate
phenols	Mix with neutral iron(III) chloride.	Purple colour forms.

Reactions of Metal Ions

Ion	Cu^{2+}	Fe^{2+}	Co^{2+}	Al^{3+}	Fe^{3+}	Cr^{3+}
aqueous ion	blue solution	pale green solution	pink solution	colourless solution	brown solution†	grey-green solution†
add OH^-	blue precipitate	pale green precipitate	blue precipitate	white precipitate	brown precipitate	green precipitate
excess OH^-	blue precipitate	pale green precipitate*	blue precipitate	colourless solution	brown precipitate	green solution
add NH_3	blue precipitate	pale green precipitate	blue precipitate*	white precipitate	brown precipitate	green precipitate
excess NH_3	dissolves to dark blue solution	does not dissolve	dissolves to straw coloured solution	does not dissolve	does not dissolve	dissolves to purple solution
add CO_3^{2-}	green precipitate	green precipitate	pink precipitate	CO_2 evolved white precipitate	CO_2 evolved brown precipitate	CO_2 evolved green precipitate
add excess HCl	yellow solution		blue solution			

* These are oxidised by oxygen in the air to the 3+ ion with a brown colour.
† This colour is a result of hydrolysis.

Organic Pathways

The central organic reactions are represented. Check individual chapters for interconversion of other haloalkanes, diols and nitriles.

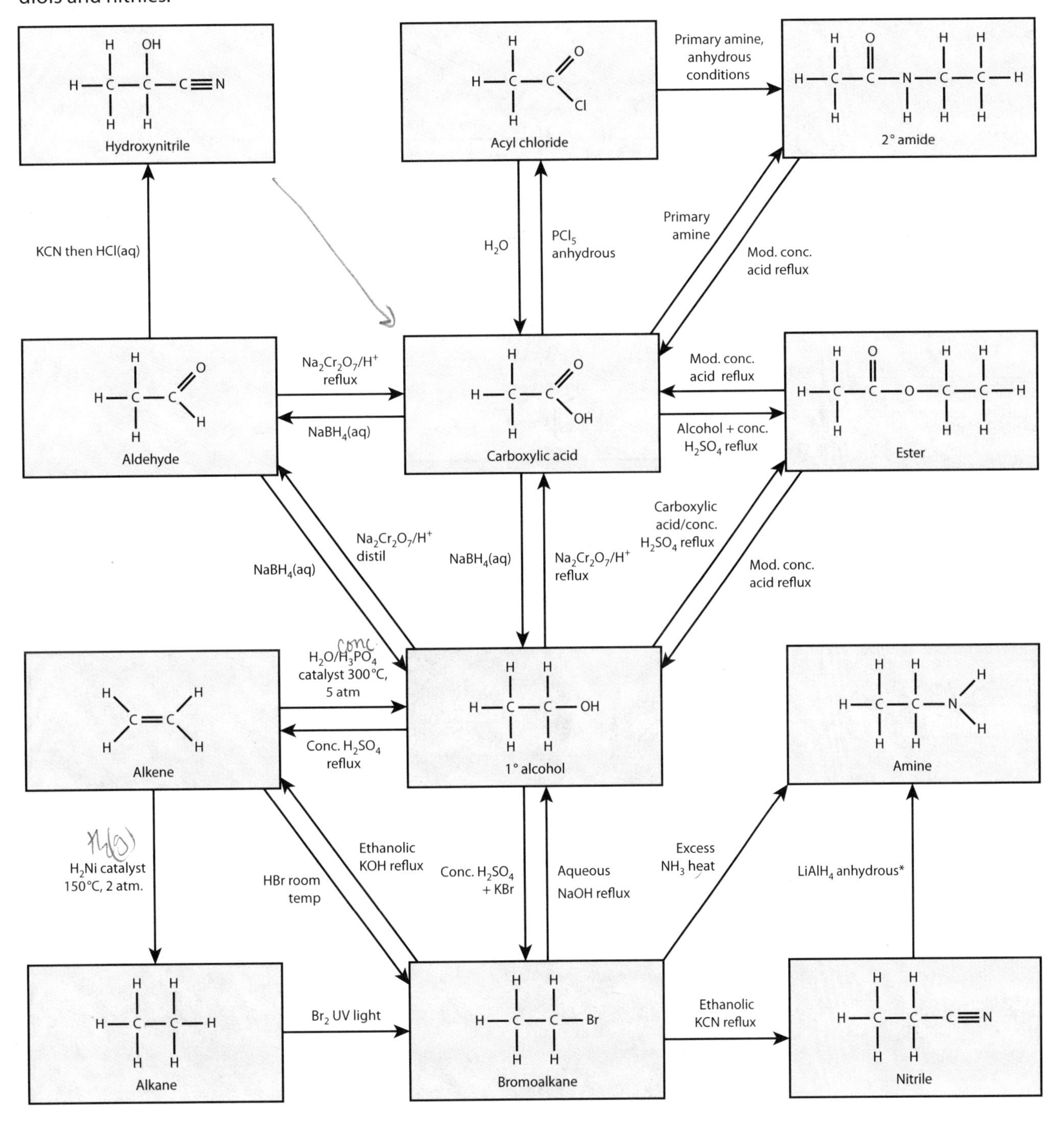

Note: This represents one way in which these conversions can be made.

* Makes the amine with one more carbon atom: propylamine not ethylamine as shown in diagram.

Reactions of Aromatic Compounds

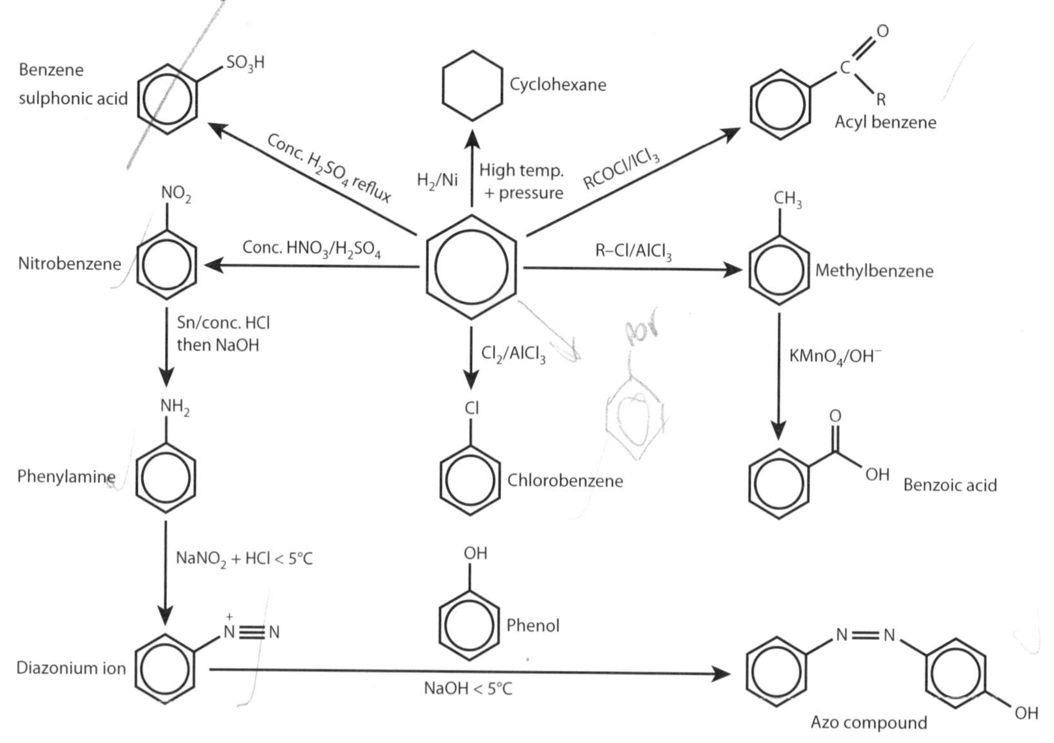

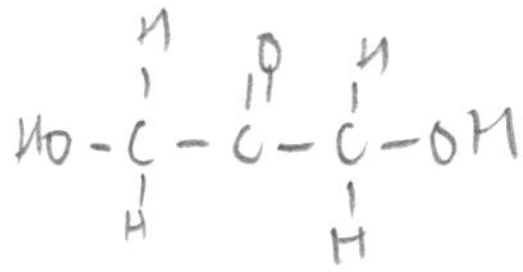

PRACTICE QUESTIONS

1. This question is about dihydroxyacetone, $CH_2(OH)COCH_2OH$. Which is correct for this molecule? **[1 mark]**

	Brick red precipitate with Fehling's solution	Silver mirror with Tollens' reagent
A	yes	yes
B	yes	no
C	no	yes
D	no	no

2. A solution of metal ions gives a green precipitate with sodium hydroxide and does not dissolve when excess is added. The metal ion is **[1 mark]**

A Cu^{2+}

B Fe^{2+}

C Cr^{2+}

D Al^{3+}

3. 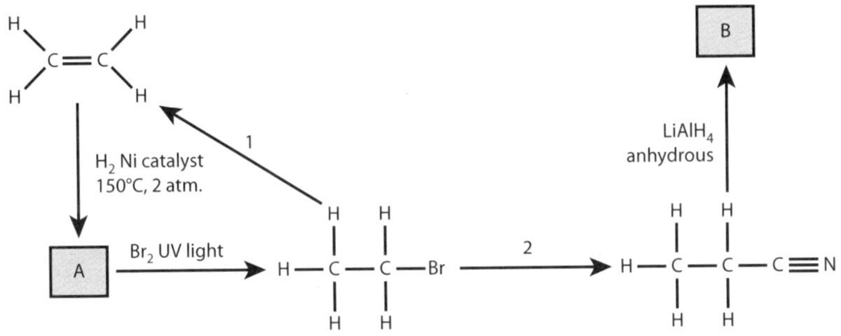

a) Draw the structural formula of molecule A. **[1 mark]**

b) Why is UV light a reaction condition in converting A to a bromoalkane? **[1 mark]**

c) What are the reagents and conditions for reaction 1? **[2 marks]**

d) What type of reaction mechanism does reaction 1 use? **[1 mark]**

e) Name molecule B. **[1 mark]**

f) Give the reactants and conditions for reaction 2. **[2 marks]**

4. **a)** Complete the reaction sequence below. **[2 marks]**

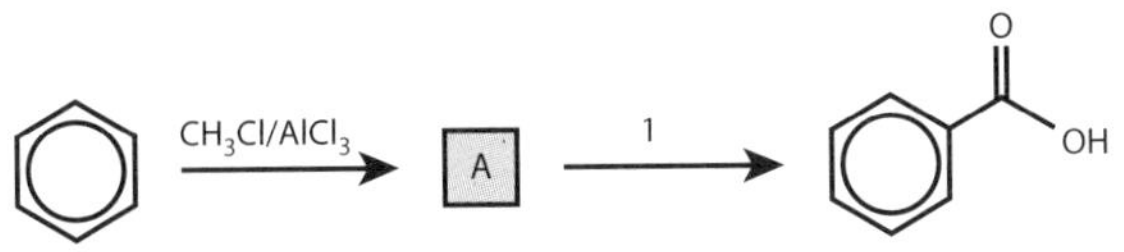

b) Describe a three-step reaction pathway from benzene to an azo dye. **[6 marks]**

Answers

Day 1

Equilibria

QUICK TEST (Page 6)

1. $\frac{[ClON_2(g)]}{[ClO(g)][NO_2(g)]}$
2. $mol^{-1}dm^3$ or dm^3mol^{-1}
3. 0.4 mol dm^3
4. $K_p = \frac{pHCl^2}{pH_2 \times pCl_2}$
5. No units
6. $K_{sp} = [Pb^{2+}_{(aq)}][I^-_{(aq)}]^2$

PRACTICE QUESTIONS (Page 6)

1. C **[1]**
2. D **[1]**
3. a) $K_c = \frac{[C_2H_5COCH_3][H_2O]}{[C_2H_5OH][CH_3COOH]}$ **[1]**
   b) $1 \times 24 \times 10^{-3} = 0.024$ moles acid **[1]**
   $0.024 \div 200 \times 10^{-3} = 0.12$ mol dm^{-3} **[1]**
   c)

	acid	alcohol	ester	water
Starting moles	1.2×10^{-1}	1.8×10^{-1}	0	0
Change in moles	-9.6×10^{-2}	-9.6×10^{-2}	$+9.6 \times 10^{-2}$	$+9.6 \times 10^{-2}$
Equilibrium moles	2.4×10^{-2}	8.4×10^{-2}	9.6×10^{-2}	9.6×10^{-2}

$K_c = (9.6 \times 10^{-2} \times 9.6 \times 10^{-2}) \div (2.4 \times 10^{-2} \times 8.4 \times 10^{-2}) = 4.57$
[4 marks: 1 mark for correct equilibrium moles alcohol; 1 mark for correct equilibrium moles ester – there is no need to calculate concentrations since all concentration values cancel out in the equation; 1 mark for correct calculation; 1 mark for no units added.]

4. a)

	PCl_5	PCl_3	Cl_2	
Starting moles	9.0	0.0	0.0	
Equilibrium moles	5.4	3.6	3.6	Total moles = 12.6
	[1 mark for both correct]			
Mole fraction	$5.4 \div 12.6 =$ 0.429	$3.6 \div 12.6 =$ 0.286	0.286	
Partial pressure/atm	$0.429 \times 2.63 =$ 1.13	$0.286 \times 2.63 =$ 0.752	0.752	

   b) K_p =0.50 **[1]** atm **[1]**
   [1 mark for correct calculation of mole fraction (see table above)]
   c) K_p would decrease **[1]** because the reaction is endothermic **[1]**.

5. a) $K_{sp} = [Ba^{2+}(aq)][SO_4^{2-}(aq)]$ **[1]**
   b) Solids are not included in the equilibrium expression/ the concentration of a solid does not change/solids have a concentration of 1 **[1]**.
   c) Barium sulfate would precipitate **[1]**.

Acids and pH

QUICK TEST (Page 10)

1. $HCOO^-$
2. $HClO/ ClO^-$ and H_2O/H_3O^+
3. Sulfuric acid H_2SO_4 or oxalic/ethandioic acid $C_2O_4H_2$
4. $H^+(aq) + OH^-(aq) \rightarrow H_2O(l)$
5. benzoic acid pK_a 4.2
6. 1×10^{-7} mol dm^{-3}

PRACTICE QUESTIONS (Page 11)

1. A **[1]**
2. C **[1]**
3. a) Nitric acid **[1]**
   b) It is a logarithmic value. **[1]**
   c) 6.3×10^{-5} **[1]**
   d) $C_2H_5COOH + H_2O \rightleftharpoons C_2H_5COO^- + H_3O^+$ **[1]**
   acid–base pairs: $C_2H_5COCH_3/C_2H_5COO^-$ **[1]** H_2O/H_3O^+ **[1]**
   e) $K_a = \frac{[H^+][ClO^-]}{[HClO]}$ **[2 marks: 1 mark for correct species; 1 mark for remainder correct]**
4. a) It is able to donate 3 protons. **[1]**
   b) **Either** $[OH^-] = 0.5$ mol dm^{-3} $[H^+] = 1 \times 10^{-14} \div 0.5$ **[1]**
   $[H^+] = 2 \times 10^{-14}$ $pH = -\log 2 \times 10^{-14} = 13.7$ **[1]**
   Or $[OH^-] = 0.5$ mol dm^{-3} $pOH = 0.3$ **[1]**
   $pH = 14 - 0.3 = 13.7$ **[1]**
   [Maximum 2 marks]
   c) 3 units **[1]**

Weak Acids, pH and Titrations

QUICK TEST (Page 14)

1. pH 2.45
2. 1.59×10^{-4} mol dm^{-3}
3. The concentration of acid is the same at the start and at equilibrium. The concentration of conjugate base is the same as the concentration of hydrogen ions.
4. There would be significant ionisation of the acid at equilibrium.
5. Higher than pH 7
6. Phenolphthalein, methyl red, thymol blue

PRACTICE QUESTIONS (Page 15)

1. D **[1]**
2. A **[1]**
3. a) Axes are labelled volume/pH **[1]**; the starting pH is above pH 2 **[1]**; there is a vertical region centred above pH 7 **[1]**; the final pH is above pH 10 **[1]**.

b) The anion of ethanoic acid accepts hydrogen ions from water/hydrolyses **[1]** to produce OH^- ions **[1]**
[**Also accept correct equation** $CH_3COO^- + H_2O \rightleftharpoons CH_3COOH + OH^-$]

c) pH 3 $[H^+] = 1 \times 10^{-3}$ **[1]** pK_a 4.8 = K_a 1.58×10^{-5} **[1]**
$(1 \times 10^3)^2 \div 1.58 \times 10^{-5} = 0.0633$ mol dm^{-3} **[1]**

4. a) It is a base that has accepted a proton. **[1]**

b) $K_a = \frac{[H^+][I^-]}{[HI]}$ **[1]**

c) pH 3.7 **[1]** When $\frac{[I^-]}{[HI]} = 1$ then $pK_a = pH$ **[1]**

d) **Either** the indicator would change colour at pH 3.7, which is not on the large gradient change of the pH curve and therefore before **[1]** the equivalence point **[1]** **or** it would change colour before **[1]** the vertical region of the graph **[1]**. **[Maximum 2 marks]**

5. The acid is half neutralised./moles acid = moles conjugate base **[1]**
$pK_a = pH = -\log 6.46 = 4.19$ **[1]**

Buffers

QUICK TEST (Page 18)

1. A solution that resists changes in pH when small amounts of acid or alkali are added.
2. The H^+ ions removed by the OH^- are replaced by dissociation of CH_3COOH.
3. Make a solution of a weak base and its salt.
4. pH 4
5. pH 4.2
6. H_2CO_3/HCO_3^-

PRACTICE QUESTIONS (Page 19)

1. D **[1]**
2. B **[1]**

3. a) Added H^+ reacts with HPO_4^{2-} **[1]**; the position of equilibrium moves to the left to form $H_2PO_4^-$ **[1]**. Added OH^- reacts with H^+ **[1]**. This causes the position of equilibrium to move to the right, $H_2PO_4^-$ dissociates and more H^+ is formed **[1]**.

b) $[H^+] = K_a$ [acid] ÷ [base]
$[H^+] = 6.2 \times 10^{-8} \times (0.25 \div 0.1) = 1.55 \times 10^{-7}$ **[1]**
$pH = -\log 1.55 \times 10^{-7} = 6.81$ **[1]**

c) Increase the proportion of HPO_4^{2-} **[1]**

4. a) $[H^+] = K_a$ [acid] ÷ [base] $[H^+] = 10^{-4.47} = 3.4 \times 10^{-5}$ **[1]**
$[H^+] \div K_a$ = [acid] ÷ [base] $3.4 \times 10^{-5} \div 1.7 \times 10^{-5}$ = [acid] ÷ [base] = 2 : 1 **[1]**

b) Ratio 1 salt : 2 acid, neutralise 1/3 of acid **[1]**

Moles acid = $600 \times 10^{-3} \times 0.1 = 0.06$ moles **[1]** $0.06 \times \frac{1}{3}$ = 0.02 moles NaOH required **[1]**
M_r NaOH = 40, Mass 0.02 moles = $0.02 \times 40 = 0.8$ g NaOH **[1]**

c) The ratio of base to acid would have to be very high (1 : 0.006)/there would be very little acid **[1]**. The buffer would not work well when alkali was added **[1]**.

5. a) Moles methanoic acid = $150 \times 10^{-3} \times 0.1 = 0.015$
Moles potassium hydroxide = $50 \times 10^{-3} \times 0.2 = 0.01$ **[1]**
All KOH used up in making salt, moles salt = 0.01, 0.01 moles acid used up in making salt. Remaining moles acid = 0.005 **[1]**
Total volume = 200×10^{-3}
Concentration methanoic acid = 0.005 ÷ 0.2
= 0.025 mol dm^{-3}
Concentration salt = 0.01 ÷ 0.2 = 0.05 mol dm^{-3}
$[H^+] = K_a$ [acid] ÷ [base]
$[H^+] = 1.58 \times 10^{-4} \times ([0.025] \div [0.05]) = 7.9 \times 10^{-5}$ **[1]**
$pH = -\log 7.9 \times 10^{-5}$ = pH 4.1. **[1]**
[Accept correct answer without working for full marks]

b) No change in pH **[1]**

Day 2

Lattice Enthalpy

QUICK TEST (Page 22)

1. The enthalpy change when one mole of ionic solid is formed from gaseous ions.
2. $0.5Br_2(g) \rightarrow Br(g)$
3. Energy is released when an electron joins a neutral atom.
4. Larger
5. Twice the standard enthalpy of change of formation of calcium oxide or 2 × enthalpy of combustion of calcium.
6. 0.5 × [(–atomisation enthalpy of sodium) + (enthalpy of formation) – (lattice enthalpy) – (first electron affinity of chlorine) – (first ionisation energy of sodium)

PRACTICE QUESTIONS (Page 23)

1. C **[1]**
2. D **[1]**
3. **a)** ΔH_1 = standard enthalpy of formation of CsCl **[1]**
   ΔH_2 = standard enthalpy of atomisation of Cs **[1]**
   ΔH_3 = first ionisation enthalpy of Cs **[1]**
   ΔH_4 = standard enthalpy of atomisation of Cl **[1]**
   b) It involves only bond making. **[1]**
   c) –(–364) – (+121) – (+376) – (+79) + (–433)
   = –645 **[1]** kJ mol^{-1} **[1]**
4.

$Ca^{2+}(g) + 2F(g)$

Second ionisation enthalpy of Ca — $Ca^{+}(g) + 2F(g)$ — 2 × first electron affinity of F

First ionisation enthalpy of Ca — $Ca(g) + 2F(g)$ — $Ca^{2+}(g) + 2F^{-}(g)$

2 × enthalpy of atomisation of F — $Ca(g) + F_2(g)$ — Lattice enthalpy of CaF_2

Enthalpy of atomisation of Ca — $Ca(s) + F_2(g)$

Enthalpy of formation of CaF_2 — $CaF_2(s)$

[6 marks: 3 marks for correct species at each energy level (–1 mark for each error); 3 marks for correct enthalpy changes (–1 mark for each error)]

Theoretical Lattice Enthalpies/Enthalpy of Solution

QUICK TEST (Page 26)

1. The amount of covalent nature in an ionic lattice.
2. The lattice enthalpy becomes more negative.
3. MgF_2
4. I^- is larger and more polarisable.
5. The enthalpy change when one mole of solid is dissolved in water to infinite dilution/until there is no further change in enthalpy, under standard conditions.
6. Mg^{2+}

PRACTICE QUESTIONS (Page 27)

1. A **[1]**
2. D **[1]**
3. **a)** The lattice enthalpy for LiCl is more negative than for NaCl. This means that the ionic bonding is stronger in LiCl than in NaCl **[1]**. The ionic radius of Li^+ is smaller than that of Na^+ **[1]**. This means that there is a stronger electrostatic attraction between Li^+ and Cl^- than between Na^+ and Cl^- **[1]**.
   b) The experimental lattice enthalpy is more negative than the theoretical lattice enthalpy. The theoretical value is based on the charge and radius of the ions / spherical ions and the calculated electrostatic attraction between them **[1]**. The experimental value is based on measured values from a Born–Haber cycle **[1]**. The difference shows that there is some covalent nature in the ionic lattice/the ions are distorted in shape **[1]**.
   c) AgCl difference = 17.4% **[1]** NaCl difference = 0.65% **[1]** AgCl lattice has more covalent nature than the NaCl lattice **[1]**.
4. **a)**

$BaBr_2(s)$ —$\Delta H^{\ominus}_{solution}$→ $BaBr_2(aq)$

$\Delta H^{\ominus}_{LE}$ — $Ba^{2+}(g) + 2Br^-(g)$ — $\Sigma\Delta H^{\ominus}_{hydration}$

   [2 marks: 1 mark for correct species with state symbols; 1 mark for arrows in correct direction]
   b) –(–1937) + (–1360) + 2(–307) = –37 kJ mol^{-1} **[2 marks: 1 mark for correct numerical value; 1 mark for correct sign and units, even if value incorrect]**

Entropy, S

QUICK TEST (Page 30)

1. Positive
2. Positive
3. Negative
4. Negative
5. –25 J K^{-1} mol^{-1}
6. Yes, ΔG must be negative.

PRACTICE QUESTIONS (Page 31)

1. D **[1]**
2. B **[1]**
3. **a)** Negative **[1]** because the number of moles of gas decreases **[1]**.
   b) 2(221) – 3(131) – 192 = –143 J K^{-1} mol^{-1} **[1]**

c) **Either** $\Delta S_{tot} = \Delta S_{sys} + \Delta S_{sur}$

$\Delta S_{surr} = -\left(\frac{-92000}{298}\right)$ **[1]** $\Delta S_{tot} = -143 + 309 = +166$ J K^{-1} mol^{-1} **[1]**

When ΔS_{tot} is positive the reaction is feasible **[1]**.

Or $\Delta G = \Delta H - T\Delta S$

$T\Delta S = 298 \times -143$ **[1]**

$\Delta G = -92000 - (298 \times -143) = -49386$ J mol^{-1} **[1]**

When ΔG is negative the reaction is feasible **[1]**.

[Maximum 3 marks]

d) $T = \frac{\Delta H}{\Delta S_{surr}}$ **[1]** $\frac{92000}{143} = 643$ K **[1]**

e) The conditions are not standard/the pressure is very high. **[1]**

4. a)

ΔS_{sys}	ΔH	**Feasibility**
positive	negative	feasible at all temperatures **[1]**
positive	positive	feasible at high temperature, not feasible at low temperature
negative	negative	feasible at low temperatures, not at high temperatures **[1]**
negative	positive	not feasible at any temperature **[1]**

b) $\Delta S_{sys} = 83 - 218 - 44 = -179$ J K^{-1} mol^{-1} **[1]**

$T = \frac{70000}{179} = 391$ K **[1]**

The Rate Equation

QUICK TEST (Page 34)

1. Rate = k [A][B]^2
2. Third order
3. Second order
4. × 8

PRACTICE QUESTIONS (Page 35)

1. A **[1]**
2. C **[1]**
3. a) Experiment 1 and 2 [A] remains constant, [B] doubles, rate doubles **[1]**. First order with respect to B **[1]**. Experiment 2 and 3 [B] remains constant, rate ÷ 4, [A] halves **[1]**. Second order with respect to A **[1]**.
   b) 0.30×10^{-3} mol dm^{-3} s^{-1} **[1]**
   c) Rate = k[A]^2[B] **[1]**
   d) 0.15 **[1]** mol^{-2} dm^6 s^{-1} **[1]**
4. a) It will increase **[1]**.
   b) 4.0×10^{-4} mol dm^{-3} s^{-1} **[1]**
   c) 1.6×10^{-3} mol dm^{-3} s^{-1} **[1]**
   d) 3.2×10^{-3} mol dm^{-3} s^{-1} **[1]**

Day 3

Experiments to Find Orders of Reaction

QUICK TEST (Page 38)

1. Measure volume of CO_2 against time / measure mass loss against time/monitor pH / titrate acid content.
2. A horizontal line, y-axis = rate, x-axis = concentration (see page 36).
3. The time taken for the concentration to halve.
4. A curve with an increasing half-life.
5. The gradient is the rate constant.
6. The rate of reaction for the same concentrations at different temperatures.

PRACTICE QUESTIONS (Page 39)

1. C **[1]**
2. A **[1]**
3. a) Iodine is coloured. **[1]**
   b) Choose five different concentrations of iodine **[1]**. Add an excess of propanone (and acid) **[1]**. Follow the course of the reaction/loss of iodine with time for each concentration **[1]**. Keep the temperature the same for each experiment **[1]**.
   c) The reaction is zero order with respect to iodine **[1]**.
   d)

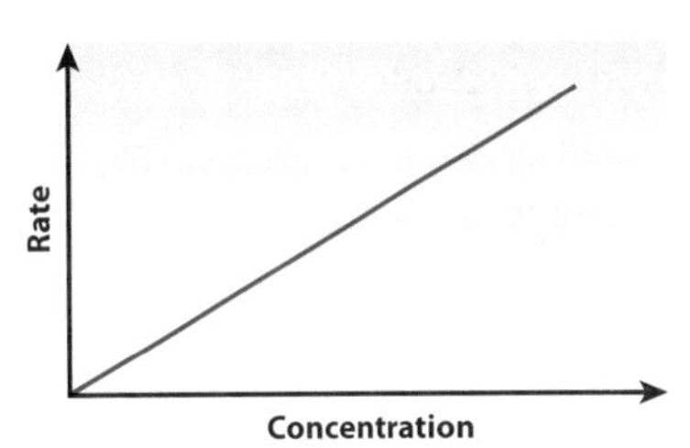

   A straight line passing through the origin **[1]**.
   e) Plot a graph of Rate vs [H$^+$]^2 **[1]**. A straight line confirms second order **[1]**.
   f) Rate = k[CH_3COCH_3][H$^+$]^2 **[1]**; the reaction is third order **[1]**.
4. a) $5200 \times 8.31 = 43212$ **[1]** 43.2 kJ mol^{-1} **[1]**
   b) Extrapolate the line of the graph; where it crosses the y-axis = value of A **[1]**.

Reaction Mechanisms

QUICK TEST (Page 42)

1. The slowest step in a multistep reaction.
2. 2
3. S_N1
4. $[HO\cdots C(CH_3)(H)(H)\cdots Cl]^-$

5. It is not possible to tell without experimental values.
6. 1/time taken for the colour to appear or 0.5[thiosulfate]/ time taken for colour to appear.

PRACTICE QUESTIONS (Page 43)

1. B **[1]**
2. D **[1]**
3. **a)** $(CH_3)_2C{=}O + I_2 \rightarrow CH_3COCH_2I + HI$ **[1]**
   b) It is a catalyst **[1]**; it is required for step 1 and regenerated in step 2 **[1]**.
   c) The slowest step **[1]**.
   d) Step 3 **[1]**; it requires I_2, which does not appear in the rate equation **[1]**.
4. **a)** There are three reactant particles **[1]**. There is a low probability that these would collide **[1]**.
   b) Any mechanism where there is a step using O_3 and NO_2 **[1]** and where the reactants and products add up to the reaction equation **[1]**
   e.g. step 1 $O_3 + NO_2 \rightarrow O_2 + NO_3$
   step 2 $NO_3 + NO_2 \rightarrow N_2O_5$
   c) The step that includes O_3 and NO_2 as reactants **[1]**.

Redox

QUICK TEST (Page 46)

1. Ag^+
2. Mg
3. The standard hydrogen electrode
4. Ag^+

PRACTICE QUESTIONS (Page 47)

1. C **[1]**
2. D **[1]**
3. **a)** The voltage/e.m.f./E_{cell} **[1]** when a half-cell is connected to a standard hydrogen electrode **[1]** under standard conditions **[1]**.
   b) Cr is a better reducing agent than H_2 but H_2 is a better reducing agent than Cu **or** Cu^{2+} is a better oxidising agent than H^+ but H^+ is a better oxidising agent than Cr^{3+} **[1]**.
   c) Hg^{2+} **[1]**
   d) Cr **[1]** and Sn **[1]**
   e) $Cr|Cr^{3+}||Cu^{2+}|Cu$ **[2 marks: 1 mark for correct left-hand side of cell; 1 mark for correct right-hand side]**
   f) $0.34 - (-0.74) = 1.08$ V **[1]**
4. **a)**

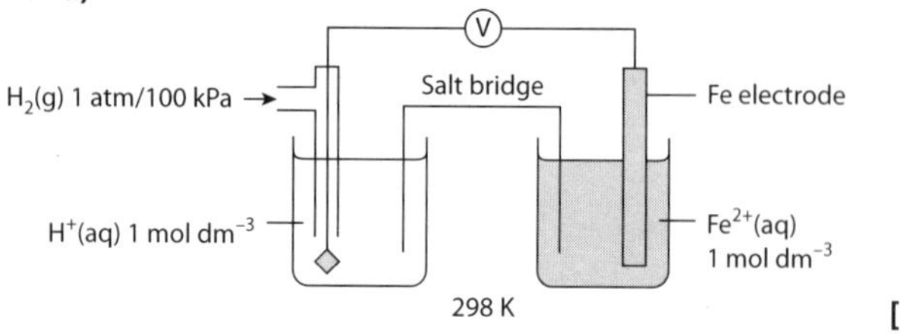

[5]

   b) From the Fe^{2+}/Fe half-cell to the hydrogen electrode **[1]**
   c) $2H^+(aq) + 2e^- \rightarrow H_2(g)$ **[1]**
   d) $H_2(g)$ **[1]**

Redox Equations

QUICK TEST (Page 50)

1. The conditions are not standard/the reaction is very slow.
2. $Zn + 2V^{3+} \rightarrow Zn^{2+} + 2V^{2+}$
3. $NO_3^- + 4H^+ + 3e^- \rightarrow NO + 2H_2O$
4. No
5. Green
6. It is negative.

PRACTICE QUESTIONS (Page 51)

1. B **[1]**
2. C **[1]**
3. **a)** NO_3^- has a more positive $E^\ominus$ than Sn **[1]**. NO_3^- oxidises Sn to Sn^{2+} / Sn reduces NO_3^- to NO_2^- **[1]**. Sn^{2+} is soluble **[1]**.
   b) $Sn(s) + NO_3^-(aq) + 2H^+(aq) \rightarrow Sn^{2+}(aq) + NO_2^-(aq) + H_2O(l)$ **[1]**
   c) -0.14 volts **[1]**
   d) The conditions are not standard / the reaction is very slow **[1]**.
   e) There is a large difference in $E^\ominus$ between Fe and H_2/Fe and NO_3 **[1]**.
   f) $NO_2^- + 7H^+ + 6e^- \rightarrow NH_3 + 2H_2O$ **[2 marks: 1 mark for correct species in reactants and products; 1 mark for correct number of H+ and electrons]**
4. **a)** +4 **[1]**
   b) $2I_2(aq) + 2V^{2+}(aq) + 2H_2O(l) \rightarrow 2VO^{2+}(aq) + 4H^+(aq) + 4I^-(aq)$ **[2 marks: 1 mark for correct species in reactants and products; 1 mark for balanced equation]**

Day 4

Storing Electricity/Rusting

QUICK TEST (Page 54)

1. The chemicals are used up.
2. $Cd(OH)_2 + 2e^- \rightarrow Cd + 2OH^-$
3. Cadmium is a toxic metal.
4. Lithium cells are rechargeable; zinc alkaline batteries are not.
5. $2H_2 + O_2 \rightarrow 2H_2O$
6. It is used in the reduction reaction $O_2 + 2H_2O + 4e^- \rightarrow 4OH^-$

PRACTICE QUESTIONS (Page 55)

1. D [1]
2. B [1]
3. a) $CH_3OH + 1.5O_2 \rightarrow CO_2 + 2H_2O$ [1]
   b) $CH_3OH + H_2O \rightarrow CO_2 + 6H^+ + 6e^-$ **[2 marks: 1 mark for H^+ in products; 1 mark for completely correct]**
   c) **Any two from:** methanol is toxic, ethanol is not; methanol is more costly to produce; ethanol is easy to produce **[2 marks: 1 mark for each point made]**
   d) Production of methanol requires an energy input / catalysts require high energy costs to produce **[1]**; methanol releases CO_2 to the environment **[1]**.
4. a) $Zn \rightarrow Zn^{2+} + 2e^-$ **[1]**
   b) $Fe^{2+} + 2e^- \rightarrow Fe$ Fe^{2+} is reduced to Fe **[1]** by electrons from the zinc **[1]**. Diagram shows electrons moving through the metal of the hull from zinc to point of rusting **[1]**.
   c) The iron/steel around the copper would corrode more rapidly **[1]**. The $E^\ominus$ of Cu is more positive than Fe so the copper plate will oxidise the Fe **[1]**.

d-block Elements

QUICK TEST (Page 58)

1. Cr [Ar] $3d^5$ $4s^1$ Cr^{3+} [Ar] $3d^3$
2. The catalyst is in the same physical state as the reactants.
3. Pt/Rh in a car catalytic converter. / Ni in the hydrogenation of vegetable oils. / Fe in the manufacture of NH_3 by the Haber process. / MnO_2 in the decomposition of H_2O_2.
4. They can oxidise one reactant and reduce the other.
5. A reaction product acts as a catalyst.
6. Iodine/potassium manganate/potassium dichromate

PRACTICE QUESTIONS (Page 59)

1. C **[1]**
2. B **[1]**
3. a) The reactants are in a different phase to the catalyst **[1]**.
   b) Reactants adsorb to the catalyst surface **[1]**. Bonds in the reactants break **[1]**. New bonds form in the products **[1]**. The product desorbs and diffuses away from the catalyst **[1]**.
   c) They have available d and s orbital electrons, which can form weak bonds with reactants **[1]**.
4. a) $2Fe^{3+} + 2I^- \rightarrow 2Fe^{2+} + I_2$ **[1]**
   $2Fe^{2+} + S_2O_8^{2-} \rightarrow 2Fe^{3+} + 2SO_4^{2-}$ **[1]**
   Both reactants are converted to product and the catalyst is regenerated (or overall reaction equation) **[1]**.
   b) The first step would be the reduction of $S_2O_8^{2-}$ and the second would be the oxidation of iodide **[1]**.
   c) Zn does not show variable oxidation states in its ion **[1]**. so will not both oxidise and reduce reactants **[1]**
5. Moles of $S_2O_3^{2-} = 31.45 \times 0.015 = 4.72 \times 10^{-4}$ **[1]**
   Moles copper $= 4.72 \times 10^{-4}$ in sample **[1]**
   Moles copper in whole brass sample $= 4.72 \times 10^{-3}$ **[1]**
   Mass copper $= 4.72 \times 10^{-3} \times 63.5 = 0.30$ g **[1]**
   Percentage copper in brass $= (0.30 \div 0.5) \times 100 = 60\%$ **[1]**

Complex Ions

QUICK TEST (Page 62)

1. 6
2. A ligand that can form one coordinate bond.
3. Tetrahedral
4. It results in an increase in entropy.
5. Acid–base
6. Blue

PRACTICE QUESTIONS (Page 63)

1. C **[1]**
2. D **[1]**
3. a) Ligand exchange/substitution **[1]**
   b) It has not changed **[1]**
   c) $[Fe(H_2O)_5(SCN)]^{2+}$ (structure: Fe centre with H_2O, H_2O, OH_2, H_2O, H_2O and SCN ligands; charge 2+)
   Octahedral
   [3 marks: 1 mark for correct 3-D shape and charge; 1 mark for bonds attached to the correct atoms SCN can be in any position; 1 mark for naming shape]
   d) The ligands surrounding the Fe^{3+} cause the 3d orbitals to split into different energies **[1]**. The difference in energy between the two levels, ΔE, is the same as the energy of photons in the visible spectrum **[1]**. $E = hf$ **[1]** Electrons in the lower energy d orbital absorb light and are excited to the higher energy orbitals **[1]**. This causes the complementary colour of light to be seen **[1]**. Changing the ligands changes the value of ΔE (and so the frequency of light absorbed) **[1]**.
4. a) A ligand that can form two coordinate bonds. **[1]**

b)

Accept

[2 marks: 1 mark for correct 3-D shape; 1 mark for coordinate bonds attached to N atom]

c) The complex can have both water molecules next to each other, cis isomer **[1]** or on either side of the molecule, trans isomer **[1]**.

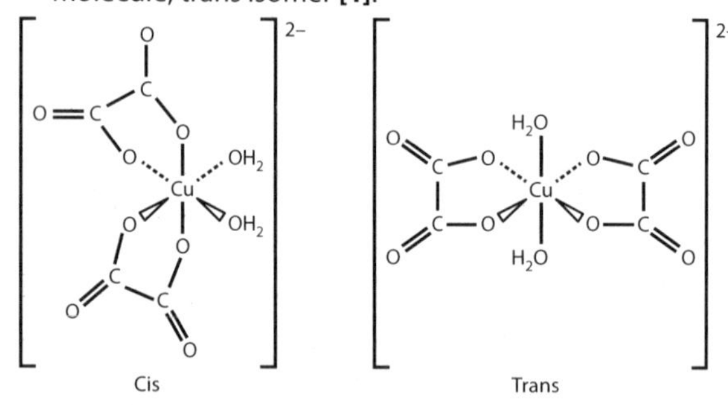

Cis Trans

[1 mark for 3-D diagram; 1 mark if correctly labelled]

d) $[Cu(C_2O_4)_2(H_2O)_2]^{2-} + EDTA \rightleftharpoons [CuEDTA]^{4-} + 2C_2O_4^{2-} + 2H_2O$

e) EDTA displaces four molecules and so causes an increase in entropy **[1]**.

Period 3 Elements and Their Oxides

QUICK TEST (Page 66)

1. $P_4(s) + 5O_2(g) \rightarrow P_4O_{10}(s)$
2. No reaction, it is insoluble.
3. $SO_3 + 2NaOH \rightarrow Na_2SO_4 + H_2O$
4. Contains a dative covalent bond between two molecules.
5. The tendency for a Group 4 atom not to use s orbital electrons in making bonds.
6. $SiCl_4 + 4H_2O \rightarrow SiO_2 + 4HCl$

PRACTICE QUESTIONS (Page 67)

1. C **[1]**
2. C **[1]**
3. **a)** Ge $1s^2\ 2s^2\ 2p^6\ 3s^2\ 3p^6\ 3d^{10}\ 4s^2\ 4p^2$ **[1]**

   b) CO_2/CH_4/any carbon compound with 4 covalent bonds **[1]**
   $PbCl_4$/PbO_2/any Pb compound where Pb has combining number of 4 **[1]**

   c) Pb has 2 electrons in an s orbital and 2 electrons in a p orbital **[1]**. In the +2 state it only uses the p orbital electrons **[1]**.

   d) They are less stable than compounds in the +4 state **[1]**.

4. **a)** Amphoteric **[1]**

   b) $Al_2O_3 + 2NaOH + 3H_2O \rightarrow 2NaAl(OH)_4$ **[1]**
   $Al_2O_3 + 6HCl \rightarrow 2AlCl_3 + 3H_2O$ **[1]**

   c)

   [1 mark – arrows are not required for the mark]

5. **a)** N B

   The dot and cross can be reversed or both atoms can use dots or crosses **[1]**.

   b) The nitrogen donates its lone pair of electrons to form a dative covalent bond **[1]**.

   c) Boron nitride can form two **giant covalent** structures **[1]**: hexagonal boron nitride forms layers of hexagons with weak intermolecular bonding between layers/suitable diagram **[1]**; the planes/layers are able to slide over each other making hexagonal boron nitride a good lubricant **[1]**; cubic boron nitride forms tetrahedral units similar to diamond, with alternating boron and nitrogen atoms/suitable diagram **[1]**; four covalent bonds to each atom makes a hard structure that can be used as an abrasive **[1]**; see diagram pg 65.

Day 5

Carbonyl Compounds

QUICK TEST (Page 70)

1. Pentan-3-one

```
      H   H   O   H   H
      |   |   ‖   |   |
  H — C — C — C — C — C — H
      |   |       |   |
      H   H       H   H
```

2. Add a solution of 2,4-dinitrophenylhydrazine.
3. A carbonyl on the first carbon is an aldehyde.
4. Fehling's solution/copper ions and potassium hydroxide, Tollens' reagent/ammoniacal silver nitrate/acidified potassium dichromate.
5. Sodium hydridoborate/$NaBH_4$ or lithium hydridoaluminate/ $LiAlH_4$.
6. HCN with alkali/NaCN followed by heating with hydrochloric acid.

PRACTICE QUESTIONS (Page 71)

1. C **[1]**
2. B **[1]**
3. **a)** Add 2,4-DNP **[1]**; a yellow precipitate confirms a carbonyl group **[1]**.
   b) Add silver nitrate in ammonia/Tollens' reagent – silver mirror confirms aldehyde **or** copper ions and potassium hydroxide/Fehling's solution, heat – brick-red precipitate confirms aldehyde **[2 marks: 1 mark for test; 1 mark for outcome]**
   c) Take the melting point of the 2,4-DNP derivative/take IR spectrum **[1]** and compare with database **[1]**.
4. **a)**

```
      H   H   δ–O   H
      |   |    ‖    |
  H — C — C — C — C — H
      |   |   δ+    |
      H   H         H
```

[2 marks: 1 mark for correct structure; 1 mark for correct partial charges]

b) Butan-2-ol **[1]**; acidified potassium dichromate **[1]**; reflux **[1]**

c)

```
                              H–O–H
  H  H δ–O  H            H  H  O:  H            H  H  OH  H
  |  |δ+‖   |            |  |  |   |            |  |  |   |
H–C–C–C–C–H  →  H–C–C–C–C–H  →  H–C–C–C–C–H
  |  |  :H–  |            |  |  |   |            |  |  |   |
  H  H       H            H  H  H   H            H  H  H   H
```

[3 marks: 1 mark first two arrows correct; 1 mark second two arrows correct; 1 mark correct structures]

d)

```
                           H+
  H  H δ–O  H            H  H  O–  H            H  H  OH  H
  |  |δ+‖   |            |  |  |   |            |  |  |   |
H–C–C–C–C–H  →  H–C–C–C–C–H  →  H–C–C–C–C–H
  |  | :CN– |            |  |  |   |            |  |  |   |
  H  H      H            H  H  CN  H            H  H  CN  H
```

[4 marks: 1 mark for correct dipoles; 1 mark for first two arrows correct; 1 mark for second two arrows correct; 1 mark for correct structures]

e) The reaction produces a chiral centre in the molecule **[1]**. The molecule is planar at the carbonyl group **[1]** so the CN^- is equally likely to attack from either side **[1]**.
f) Warm with hydrochloric acid **[1]**.
g) 2-hydroxy-2-methylpentanoic acid **[1]**

Carboxylic Acids and Their Derivatives

QUICK TEST (Page 74)

1.

```
        O
 HO     ‖
   \/\/ \
   ‖     OH
   O
```

2. $Ca + 2CH_3COOH \rightarrow Ca(CH_3COO)_2 + H_2$
3. Reduce with $LiAlH_4$.
4. HCl
5. Butylethanoate
6. The anion of the acid is stabilised by delocalisation of electrons so the equilibrium is further to the right when compared to ethanol.

PRACTICE QUESTIONS (Page 75)

1. D **[1]**
2. A **[1]**
3. **a)** Oxidise to hexanedioic acid **[1]** by refluxing with acidified potassium dichromate **[1]**. Convert the carboxylic acid groups to acyl chlorides **[1]** by reaction with PCl_5 **[1]**.
   b)

```
      H   H       O   H   H   H   H   O       H   H
      |   |       ‖   |   |   |   |   ‖       |   |
  H — C — C — O — C — C — C — C — C — C — O — C — C — H
      |   |           |   |   |   |           |   |
      H   H           H   H   H   H           H   H
```

[2 marks: 1 mark for correct ester group; 1 mark for two ester groups]

c) Ester **[1]**
d) HCl (not hydrochloric acid) **[1]**

4. **a)** A = alcohol **[1]** B = carboxylic acid **[1]** C = phenol **[1]**
   b) A does not react **[1]**; C neutralises the base **[1]**.
   c) B reacts to produce carbon dioxide (water and a salt) **[1]**; C does not react **[1]**.
   d) **Any four from:** the O—H bond can break to release H^+; $R—COOH \rightleftharpoons RCOO^- + H^+$; The position of equilibrium is further to the right in B because the charge is spread across the COO group; No significant equilibrium in A so no acid behaviour; Some sharing of the charge with the benzene ring in C but the equilibrium is not as far to the right as in B; C is too weak an acid to react with a carbonate/will only react with a strong base. **[4 marks: 1 mark for each point made]**

Benzene

QUICK TEST (Page 78)

1.

```
        Cl
       /
   ⌬
  /
Cl
```

2. Electrophilic substitution

3. Concentrated nitric and sulfuric acid, below 55°C
4. $AlCl_3/FeCl_3$
5. It has a high carbon : hydrogen ratio, which causes incomplete combustion.
6. The —OH is not attached directly to the benzene ring.

PRACTICE QUESTIONS (Page 79)

1. A **[1]**
2. B **[1]**
3. **a)** Each has a hexagonal structure **[1]**. Benzene is planar, cyclohexane is not **[1]**. Benzene has a ring of delocalised electrons **[1]**. Cyclohexene has a C=C double bond/a π bond **[1]**.
   b) Benzene – no change to the bromine water **[1]**; cyclohexene – decolourises, brown to colourless **[1]**.
   c) Higher electron density between the carbon atoms of the π bonds in cyclohexane **[1]** polarises the Br_2 **[1]**. Lower electron density of the delocalised ring in benzene is not able to polarise bromine **[1]**.
   d) Electrophilic addition **[1]**
   e) Electrophilic substitution **[1]**
4. **a)** D **[1]**
   b) B **[1]**
   c) B **[1]**
   d) E **[1]**
   e) A **[1]**
5. All the carbon–carbon bonds in benzene are of the same length **[1]** but C—C single bonds are longer than C=C double bonds **[1]**.
   Benzene does not undergo electrophilic addition, a typical alkene reaction **[1]**.
   The enthalpy of hydrogenation of benzene is much smaller than the expected value **[1]**.

Organic Nitrogen Compounds

QUICK TEST (Page 82)

1. Amines can hydrogen bond with water.
2. 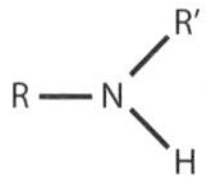 R—NHR′
3. The lone pair of electrons on nitrogen can be donated to a central metal ion.
4. Tin and concentrated hydrochloric acid.
5. Nucleophilic substitution
6. The part of a molecule responsible for its colour.

PRACTICE QUESTIONS (Page 83)

1. C **[1]**
2. D **[1]**
3. **a)** $C_3H_7Br + 2NH_3 \rightarrow C_3H_7NH_2 + NH_4Br$ **[1]**
   b) 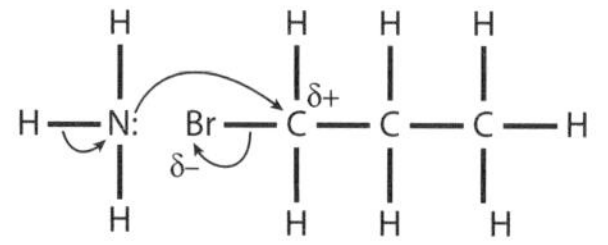 Nucleophilic substitution
   [3 marks: 1 mark for correct arrows; 1 mark for correct molecules; 1 mark for naming type of reaction]
   c) The bromoalkane can substitute again onto the amine **[1]** because there is still a lone pair of electrons on the nitrogen **[1]**; using excess ammonia **[1]**.
4. **a)** 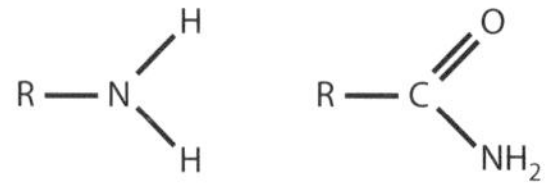 Amine — Primary amide
   [2 marks: 1 mark for each structure]
   b) 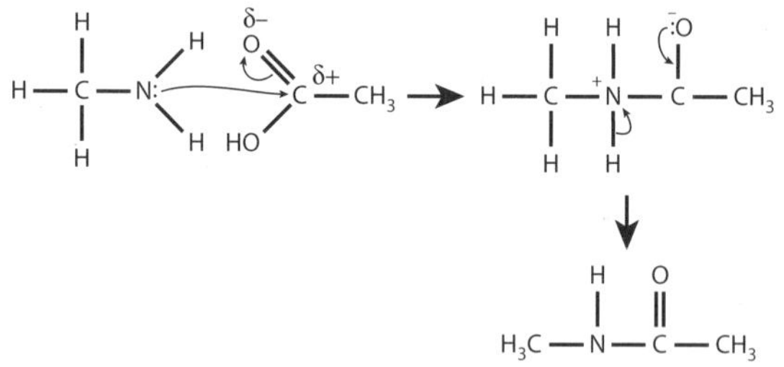 $H_3C—NH—C(=O)—CH_3$
   [3 marks: 1 mark for correct product; 1 mark for correct arrows on first step; 1 mark for correct arrows on second step]
   c) The amine will be protonated so the lone pair of electrons is not free to react **[1]**.
5. **a)** $NaNO_2$ **[1]**; HCl **[1]**; below 5°C **[1]**
   b) A coupling reaction **[1]**
   c) 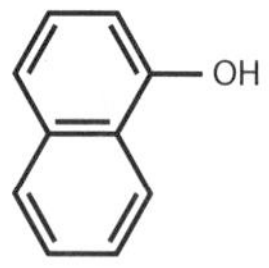 OH **[1]**

Day 6

Amino Acids and Proteins

QUICK TEST (Page 86)

1. $H_2N—CH(CH_3)—C(=O)—OH$
2. An amino acid in which the amine and acid group are both ionised.
3. It is the joining together of two molecules (amino acids) with the elimination of a small molecule (water).
4. One optical isomer
5. A substance that rotates the plane of plane-polarised light.
6. Interactions between the R groups of the amino acids, ionic bonds, hydrogen bonds, instantaneous dipole–induced dipole forces, covalent disulfide bonds.

PRACTICE QUESTIONS (Page 87)

1. A **[1]**
2. D **[1]**
3. **a)** Chiral centre* — $(CH_3)_2CH–CH(NH_2)–COOH$ **[1]**

   b) H_2N, H, CH_3, COOH around C | HOOC, H, CH_3, NH_2 around C **[2 marks: 1 mark for 3-D structure; 1 mark for any arrangement of groups providing they are mirror images with the 2-carbon atom at the centre]**

   c) $H_3N^+–CH(R)–COO^-$ **[2 marks: 1 mark for correct general structure; 1 mark for both amine and carboxyl group ionised]**

   d) Shine plane-polarised light through a sample **[1]**; one isomer rotates clockwise and the other anticlockwise/note the direction of rotation **[1]**.

   e) A 50/50 mixture of optical isomers **[1]**.
4. **a)** HO–C(=O)–CH(R′)–N(H)H + HO–C(=O)–CH(R″)–NH₂ (H HO circled) → $H_2N–CH(R′)–C(=O)–N(H)–CH(R″)–C(=O)OH$

   [2 marks: 1 mark for correct amino acid structures; 1 mark for correct peptide link]

   b) The sequence of amino acids **[1]**.

   c) Hydrogen bonding **[1]**

   d) The primary sequence determines the position of the R groups **[1]**. The bonding between the R groups holds the shape of the protein together **[1]**. Use of 'tertiary structure' or description of some of the bonding between R groups **[1]**.

Enzymes and DNA

QUICK TEST (Page 90)

1. Enzyme + substrate ⇌ enzyme substrate complex ⇌ enzyme product complex ⇌ enzyme + product
2. The substrate must be able to bind to the active site. This requires the ionisable R groups to be in the right state.
3. The part of a molecule responsible for its pharmacological properties.
4. Sugar (ribose or deoxyribose) phosphate and base
5. Coordinate bonding
6. Two

PRACTICE QUESTIONS (Page 91)

1. C **[1]**
2. B **[1]**
3. **a)** The R groups from the amino acids have ionisable groups **[1]**; some are used in the active site to bind substrate **[1]**; outside of the specific pH range groups are not in the form needed to bind the substrate **[1]**.

   b) The tertiary structure/3-D shape of an enzyme/active site is essential to its function **[1]**. This is (mainly) held together by weak intermolecular forces **[1]**, which are broken at high temperatures **[1]**.

   c) They are a complementary shape to the active site **[1]** and bind but do not result in a reaction **[1]** so blocking the substrate from binding/reducing the effective concentration of enzyme **[1]**.

   d) Neuroreceptor–neurotransmitter/hormone–cell membrane **[1]**; medicines bind to the receptor but inhibit or enhance the effect **[1]**.
4. **a)** Cytosine – Guanine (three hydrogen bonds shown)

   [2 marks: 1 mark for three bonds shown; 1 mark for all in correct position]

   b) Each base will only hydrogen bond to one other **[1]**. One strand of DNA acts as a template for a complementary strand **[1]**.

   c) Cisplatin bonds to two (guanine) bases **[1]** and prevents the complementary base from hydrogen bonding **[1]**.

   d) One amino acid **[1]**

Polymers

QUICK TEST (Page 94)

1. Permanent dipole–permanent dipole forces
2. An acidic gas is produced. The reaction has a poor atom economy.
3. It can be made from renewable resources. It has an overall lower carbon footprint than polymers made from crude oil. It is biodegradable.
4. There is hydrogen bonding between the polymer chains, which is stronger than instantaneous dipole–induced dipole forces.
5. There are no polar bonds, only strong carbon–carbon bonds holding the monomers together.
6. Production of acidic or toxic gases

PRACTICE QUESTIONS (Page 95)

1. C **[1]**
2. A **[1]**
3. **a)**

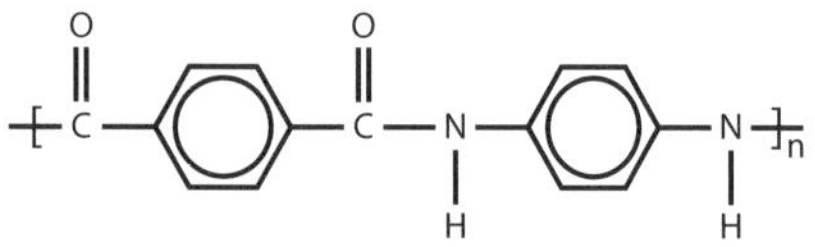

   [3 marks: 1 mark for each correct monomer structure; 1 mark correct amide link]

   b)

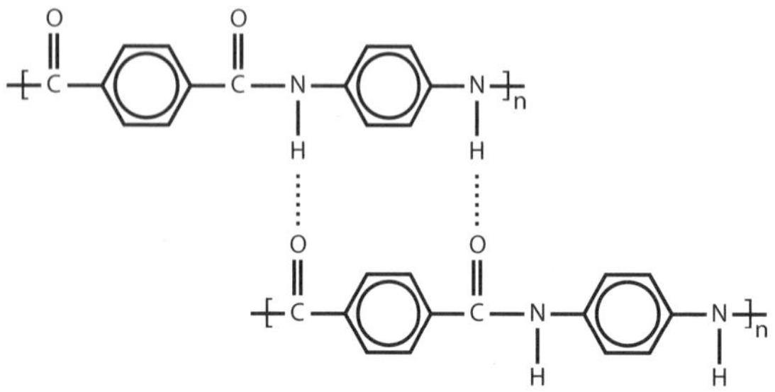

   [1]

   c)

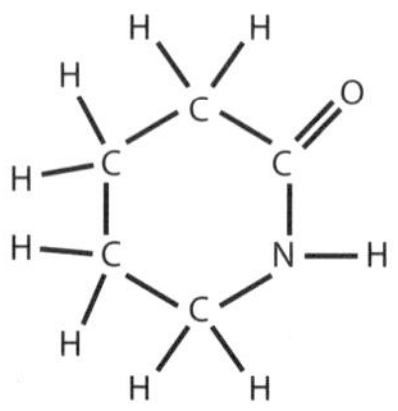

   [3 marks: 1 mark for a structure than matches the formula; 1 mark for a cyclic compound; 1 mark for the amide group]

   d) $nC_5H_9NO \rightarrow [C_5H_9NO]_n$ **[1]**

   e) No small molecule eliminated when it polymerises **[1]**.
4. **a)**

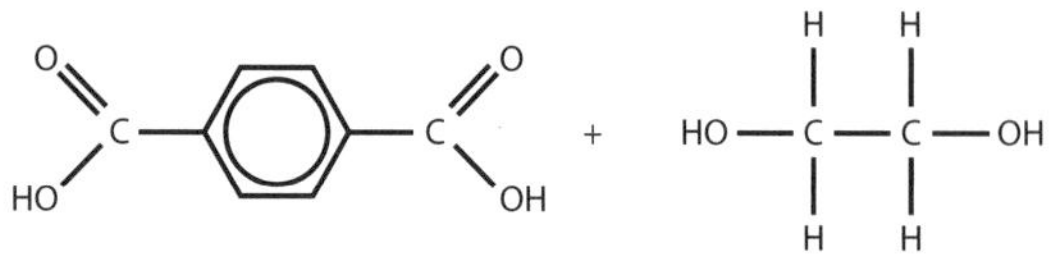

   [2 marks: 1 mark for each correct structure]

   b) Polyester **[1]**

   c) The ester group can be hydrolysed **[1]**; the polymer is biodegradable **[1]**; it is less long-lasting in the environment than other polymers **[1]**; it can be broken back down to monomers and remade **[1]**.

Organic Applications

QUICK TEST (Page 98)

1. A molecule added to a polymer to make it more flexible.
2. Ester
3. Exchanging one pairing of alcohol and carboxylic acid for another in an ester. Used in making methyl esters from oils in biodiesel.
4. By hydrolysing triglycerides with alkali
5. To attach dyes to fibres via a covalent bond
6. Sulfonate or carboxylate

PRACTICE QUESTIONS (Page 99)

1. B **[1]**
2. C **[1]**
3. **a)**

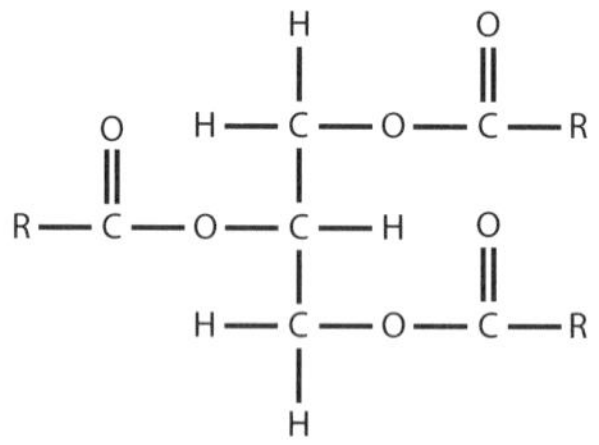

   [2 marks: 1 mark for correct 3-carbon structure for propan-1,2,3-ol; 1 mark for correct ester structure (allow any orientation of the carboxylic acid groups)]

   b) Cis double bonds introduce a kink in the molecule **[1]**; this reduces the strength of intermolecular bonding in the fat **[1]** and lowers the melting point **[1]**.

   c) Alkali/KOH/NaOH **[1]**

   d) + $3OH^-$ → + $3O^-$

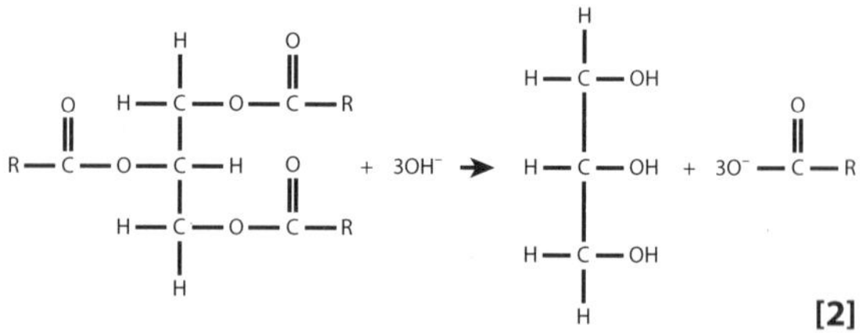

   [2]
4. **a)** OH and NH groups [OH] **[1]** allow hydrogen bonding **[1]**.

   b) $-SO_3Na$ **[1]**

   c) Either or both —N=N— groups **[1]**

   d) Electrons absorb photons/energy **[1]** from the visible spectrum **[1]** and are promoted to higher energy levels/excited **[1]**. The complementary colour is reflected/seen **[1]**.

   e) Adding groups to the chromophore makes small changes to the difference between electronic energy levels **[1]** and so a different frequency of light is absorbed **[1]**.

Day 7

Chromatography

QUICK TEST (Page 102)

1. A liquid solvent
2. A high boiling point liquid coated onto an inert solid.
3. Their volatility and solubility in the solvent
4. Running a known quantity of a known substance through the system to calibrate the machine for that substance.
5. To provide a saturated atmosphere, which prevents evaporation of solvent from the chromatogram.
6. $R_f = \frac{\text{distance moved by spot}}{\text{distance moved by solvent front}}$

PRACTICE QUESTIONS (Page 103)

1. C **[1]**
2. C **[1]**
3. **a)** Retention time increases with increasing molecular mass **[1]**.
   b) By running known samples through the machine **[1]** and measuring their retention times **[1]**.
   c) Run an external standard/known amount of benzoic acid **[1]** and calculate the relationship between peak area and mass for benzoic acid **[1]**. Measure the area under the peak for a known mass of sample **[1]**.
   d) GLC is only suitable for volatile substances **[1]**; long chain fatty acids are not volatile **[1]**. Their methyl esters are more volatile because they cannot hydrogen bond **[1]**.
   e) A mass spectrometer directly linked to the GLC **[1]**.
4. The hydrolysed mixture is spotted onto chromatography paper about 1 cm from the bottom. Known amino acid samples are spotted in separate places along the same line **[1]**. The paper is hung or stood in a developing tank containing solvent that is below the level of the sample spot and covered **[1]**. When the solvent is near the top, the position of the solvent front is marked. The plate is removed and dried **[1]**. The paper is sprayed with ninhydrin (and heated) to locate the spots **[1]**. The R_f value of spots is calculated and the R_f of spots from the hydrolysed sample is compared with the known amino acids **[1]**.

Carbon-13 NMR Spectroscopy/Mass Spectrometry

QUICK TEST (Page 106)

1. The type of carbon environments
2. Two
3. No information
4. Different isomers have different numbers of carbon environments.
5. Positive ions
6. $[C_6H_5]^+$

PRACTICE QUESTIONS (Page 107)

1. A **[1]**
2. D **[1]**
3. **a)** Three peaks **[1]**; one peak below 100 ppm, two peaks between 110 and 165 ppm **[1]**.
   b) Molecule D **[1]**; it has six carbon environments while B has seven and C has eight **[1]**.
4. **a)** 122 **[1]**
   b) $[C_6H_5]^+$ **[1]**
   c) $CO_2H\cdot$ **[1]**
   d) O, OH **[1]**

Proton NMR Spectroscopy

QUICK TEST (Page 110)

1. Because it gives no signal and does not confuse the signal from the molecule under study.
2. Tetramethylsilane/$(CH_3)_4Si$
3. The number of peaks in the spectrum = the number of proton environments.
4. The area under the peak = the relative number of protons.
5. The shift value gives the type of proton environment.
6. There are three hydrogens on the adjacent carbon atom.

PRACTICE QUESTIONS (Page 111)

1. D **[1]**
2. C **[1]**
3. **a)** 3 **[1]**
   b) The signal is split into 7 because there are 6 hydrogens on the adjacent carbon **[1]**.
   c) There are three times as many protons in this environment as there are on carbon 4 **[1]**.
   d) 2.2 (or any value in the range 2.0–2.9) **[1]**
   e) Three peaks, a doublet at around 1 **[1]** with a peak area of 6/larger than all other peaks shown **[1]**; multiplet at around 2.4, area of 1/smaller than the other peaks **[1]** and a singlet at around 2.2 **[1]**.
4.

Chemical shift	Type of environment	Number of protons	Splitting pattern	Number of protons on adjacent carbon
1.1	R—CH_3	3	triplet	2
2.4	R—CH_2—C(=O)—	2	quartet	3
3.7	—O—CH_3	3	singlet	0

[5 marks: 1 mark for each peak that has the correct assignment of environment and the correct protons on the adjacent carbon (can be shown in a table or in words) (maximum of 3 marks); 1 mark for correct structure; 1 mark for clear linking of structure with peaks]

Hydrogens on carbon 1 gives peak at 3.7

Hydrogens on carbon 3 gives the peak at 2.4

Hydrogens on carbon 4 give the peak at 1.1

Molecular formula must include 2 × oxygen since it is an ester.

Summary of Reactions/Organic Pathways

PRACTICE QUESTIONS (Page 115)

1. D **[1]**
2. B **[1]**
3. **a)**

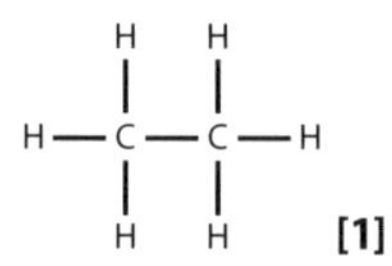

[1]

b) To initiate/form a bromine radical **[1]**

c) Ethanolic **[1]**; KOH **[1]** (or in solution with ethanol)

d) Elimination **[1]**

e) Propylamine **[1]**

f) (Ethanolic) KCN **[1]**; reflux **[1]**

4. **a)** A = methylbenzene **[1]**; 1 = alkaline $KMnO_4$ **[1]**

b) Add concentrated HNO_3 with concentrated H_2SO_4 below 55°C **[1]** to form nitrobenzene **[1]**. Reduce with Sn and concentrated HCl followed by NaOH **[1]** to make phenylamine **[1]**. Cool to below 5°C and add $NaNO_3$ and HCl **[1]**. Add phenol/phenylamine and NaOH **[1]**.

The Periodic Table

Key: relative atomic mass 1.0 — atomic symbol H — atomic number 1 — name Hydrogen

1	2											3	4	5	6	7	0
																	4.0 **He** 2 Helium
6.9 **Li** 3 Lithium	9.0 **Be** 4 Beryllium											10.8 **B** 5 Boron	12.0 **C** 6 Carbon	14.0 **N** 7 Nitrogen	16.0 **O** 8 Oxygen	19.0 **F** 9 Fluorine	20.2 **Ne** 10 Neon
23.0 **Na** 11 Sodium	24.3 **Mg** 12 Magnesium											27.0 **Al** 13 Aluminium	28.1 **Si** 14 Silicon	31.0 **P** 15 Phosphorus	32.1 **S** 16 Sulfur	35.5 **Cl** 17 Chlorine	39.9 **Ar** 18 Argon
39.1 **K** 19 Potassium	40.1 **Ca** 20 Calcium	45.0 **Sc** 21 Scandium	47.9 **Ti** 22 Titanium	50.9 **V** 23 Vanadium	52.0 **Cr** 24 Chromium	54.9 **Mn** 25 Manganese	55.8 **Fe** 26 Iron	58.9 **Co** 27 Cobalt	58.7 **Ni** 28 Nickel	63.5 **Cu** 29 Copper	65.4 **Zn** 30 Zinc	69.7 **Ga** 31 Gallium	72.6 **Ge** 32 Germanium	74.9 **As** 33 Arsenic	79.0 **Se** 34 Selenium	79.9 **Br** 35 Bromine	83.8 **Kr** 36 Krypton
85.5 **Rb** 37 Rubidium	87.6 **Sr** 38 Strontium	88.9 **Y** 39 Yttrium	91.2 **Zr** 40 Zirconium	92.9 **Nb** 41 Niobium	95.9 **Mo** 42 Molybdenum	[98] **Tc** 43 Technetium	101.1 **Ru** 44 Ruthenium	102.9 **Rh** 45 Rhodium	106.4 **Pd** 46 Palladium	107.9 **Ag** 47 Silver	112.4 **Cd** 48 Cadmium	114.8 **In** 49 Indium	118.7 **Sn** 50 Tin	121.8 **Sb** 51 Antimony	127.6 **Te** 52 Tellurium	126.9 **I** 53 Iodine	131.3 **Xe** 54 Xenon
132.9 **Cs** 55 Caesium	137.3 **Ba** 56 Barium	138.9 **La*** 57 Lanthanum	178.5 **Hf** 72 Hafnium	180.9 **Ta** 73 Tantalum	183.8 **W** 74 Tungsten	186.2 **Re** 75 Rhenium	190.2 **Os** 76 Osmium	192.2 **Ir** 77 Iridium	195.1 **Pt** 78 Platinum	197.0 **Au** 79 Gold	200.6 **Hg** 80 Mercury	204.4 **Tl** 81 Thallium	207.2 **Pb** 82 Lead	209.0 **Bi** 83 Bismuth	[209] **Po** 84 Polonium	[210] **At** 85 Astatine	[222] **Rn** 86 Radon
[223] **Fr** 87 Francium	[226] **Ra** 88 Radium	[227] **Ac*** 89 Actinium	[261] **Rf** 104 Rutherfordium	[262] **Db** 105 Dubnium	[266] **Sg** 106 Seaborgium	[264] **Bh** 107 Bohrium	[277] **Hs** 108 Hassium	[268] **Mt** 109 Meitnerium	[271] **Ds** 110 Darmstadtium	[272] **Rg** 111 Roentgenium							

Elements with atomic numbers 112–116 have been reported but not fully authenticated

lanthanides	140.1 **Ce** 58 Cerium	140.9 **Pr** 59 Praseodymium	144.2 **Nd** 60 Neodymium	144.9 **Pm** 61 Promethium	150.4 **Sm** 62 Samarium	152.0 **Eu** 63 Europium	157.2 **Gd** 64 Gadolinium	158.9 **Tb** 65 Terbium	162.5 **Dy** 66 Dysprosium	164.9 **Ho** 67 Holmium	167.3 **Er** 68 Erbium	168.9 **Tm** 69 Thulium	173.0 **Yb** 70 Ytterbium	175.0 **Lu** 71 Lutetium
actinides	232.0 **Th** 90 Thorium	[231] **Pa** 91 Protactinium	238.1 **U** 92 Uranium	[237] **Np** 93 Neptunium	[242] **Pu** 94 Plutonium	[243] **Am** 95 Americium	[247] **Cm** 96 Curium	[245] **Bk** 97 Berkelium	[251] **Cf** 98 Californium	[254] **Es** 99 Einsteinium	[253] **Fm** 100 Fermium	[256] **Md** 101 Mendelevium	[254] **No** 102 Nobelium	[257] **Lr** 103 Lawrencium

Notes

Notes

Index

Acknowledgements

The author and publisher are grateful to the copyright holders for permission to use quoted materials and images.

Cover & P1: © Shutterstock.com/Leszek Glasner

All other images are © Shutterstock.com and © HarperCollins*Publishers* Ltd

Every effort has been made to trace copyright holders and obtain their permission for the use of copyright material. The author and publisher will gladly receive information enabling them to rectify any error or omission in subsequent editions. All facts are correct at time of going to press.

Published by Letts Educational
An imprint of HarperCollins*Publishers*
1 London Bridge Street
London SE1 9GF

ISBN: 9780008179083

First published 2016

10 9 8 7 6 5 4 3 2 1

British Library Cataloguing in Publication Data.
A CIP record of this book is available from the British Library.

Series Concept and Development: Emily Linnett and Katherine Wilkinson
Commissioning and Series Editor: Chantal Addy
Author: Alison Dennis
Project Manager and Editorial: Tanya Solomons
Cover Design: Paul Oates
Inside Concept Design: Paul Oates and Ian Wrigley
Index: Simon Yapp
Text Design, Layout and Artwork: Q2A Media
Production: Lyndsey Rogers and Paul Harding
Printed in Italy by Grafica Veneta SpA